INTERNATIONAL SCIENCE AND NATIONAL SCIENTIFIC IDENTITY

AUSTRALASIAN STUDIES
IN HISTORY AND PHILOSOPHY OF SCIENCE

VOLUME 9

The titles published in this series are listed at the end of this volume.

INTERNATIONAL SCIENCE AND NATIONAL SCIENTIFIC IDENTITY

Australia between Britain and America

Edited by

R. W. HOME

University of Melbourne

and

SALLY GREGORY KOHLSTEDT

University of Minnesota

SPRINGER SCIENCE+BUSINESS MEDIA, B.V.

Library of Congress Cataloging-in-Publication Data

International science and national scientific identity : Australia between Britain and America / edited by R.W. Home and Sally Gregory Kohlstedt.
p. cm. -- (Australasian studies in history and philosophy of science ; v. 9)
Includes index.

1. Science--Australia--History--Case studies. 2. Science--Great Britain--History--Case studies. 3. Science--United States--History--Case studies. I. Home, Roderick Weir. II. Kohlstedt, Sally Gregory. III. Series.
Q127.A8I58 1991
509.94--dc20 90-5253

DOI 10.1007/978-94-011-3786-7

Printed on acid-free paper

TABLE OF CONTENTS

FOREWORD

The institutionalization of History and Philosophy of Science as a distinct field of scholarly endeavour began comparatively early – though not always under that name – in the Australasian region. An initial lecturing appointment was made at the University of Melbourne immediately after the Second World War, in 1946, and other appointments followed as the subject underwent an expansion during the 1950s and 1960s similar to that which took place in other parts of the world. Today there are major Departments at the University of Melbourne, the University of New South Wales and the University of Wollongong, and smaller groups active in many other parts of Australia and in New Zealand.

'Australasian Studies in History and Philosophy of Science' aims to provide a distinctive publication outlet for Australian and New Zealand scholars working in the general area of history, philosophy and social studies of science. Each volume comprises a group of essays on a connected theme, edited by an Australian or a New Zealander with special expertise in that particular area. Papers address general issues, however, rather than local ones; parochial topics are avoided. Furthermore, though in each volume a majority of the contributors is from Australia or New Zealand, contributions from elsewhere are by no means ruled out. Quite the reverse, in fact – they are actively encouraged wherever appropriate to the balance of the volume in question.

R. W. Home
General Editor
Australasian Studies in History and Philosophy of Science

ACKNOWLEDGEMENTS

Funding for the Conference on Nationalism and Internationalism in Science, held in Melbourne in May of 1988, was provided by the Australian government and the United States National Science Foundation under terms of the United States/Australian Co-operation Program, and by the British Council. Marc Rothenberg of the Smithsonian Institution helped with plans on the American side.

INTRODUCTION

Science, it is often said, knows no national boundaries. The operations and laws of nature are universal, and scientists from many nations have contributed to advancing our knowledge of them. When attempts have been made to reconstruct science on racial or political grounds – as 'Aryan science' in Nazi Germany, for example, or 'socialist science' by Lysenko – these have been everywhere condemned as antithetical to the very nature of the enterprise.

From another point of view, however, it is clear that science is socially embedded. Differences of language can create barriers between one scientist and another. Different cultures value science differently and provide greater or less support for scientific work. Scientists are constrained like everybody else by the forms of the particular society of which they are a part. The social institutions of science – the educational systems in which scientists receive their training, the learned societies, publication outlets, professional organizations and research institutes – vary markedly from one country to another in both structure and mode of operation.

Sociologists of knowledge argue that scientific knowledge is constructed by the social context in which it is produced. Even those who do not wish to go as far as that agree that, at the level of practice, social circumstances exert a powerful influence on scientific work. Historians of modern science have therefore increasingly turned their attention to the dynamics of scientific practice, attending more closely to the fine structure of schooling and research in different countries and historical periods. Attention to location has also pushed scholars to look beyond the traditionally recognized centres of scientific activity towards the industrial cities of England, the provinces of France and the colonies of European imperial powers.[1]

In this last case in particular, the role of social circumstances is most dramatically visible. Removed from the main European centres of scientific authority, scientists transported to colonial settings found

R.W. Home and S.G. Kohlstedt (eds.), International Science and National Scientific Identity, 1–17.

themselves dependent for both intellectual and practical support on colleagues 'at home'. They outlined their circumstances in poignant letters and sometimes in fact abandoned the colonies altogether. Moreover, this state of dependency formed in the imperial context often lasted beyond political independence, so that scientists working in the United States in the nineteenth century, for example, nevertheless long remained reliant, scientifically speaking, on the European-based leaders of their disciplines.[2]

The precise nature of the relationship of dependency that is involved in such situations has occupied the attention of a number of historians in recent years. Some have analysed the ways in which certain major centres have exercised authority over scientific communities and activities elsewhere. Others, acknowledging the visibility and powerful influence of such centres, have nonetheless challenged notions of control, depicting instead a dynamic interaction involving both competition and co-operation within networks of scientists located at both centre and periphery.

The situation is complicated by the fact that the relationship is one that changes over time. Scientific communities that were at one time dependent on a metropolitan centre such as London or Paris may at some later time be judged to be dependent no longer. Newly emerging nations may, it seems, aspire to scientific as well as political and economic independence. Indeed, George Basalla in a frequently cited paper once drew up a list of criteria by which to assess the degree to which 'scientific independence' had been achieved. These included scientists of the territory in question receiving most of their training at home rather than abroad; their work coming to enjoy local respect and financial support; their being able to find intellectual stimulation in the local scientific community, publication outlets for their work in local scientific journals, opportunities for opening 'new fields of scientific endeavor', and rewards locally bestowed for work of superior quality; and a 'proper technological base' being established in the country in question.[3]

Basalla's criteria have been criticized, especially on the grounds that he assumed that there was a unique western scientific ideology waiting to be diffused into new scientific territories, and that he treated science in isolation and so failed to take proper account of the political, economic and social forces that brought about the changes he described.[4] Moreover, whenever his model has been put to the test, it has been found not to apply to the particulars of the actual historical case. It seems that the transition from a state of scientific dependency to one of non-dependency does not follow a simple model or operate uniformly across disciplines. Nonetheless, the idea of a nation establishing a degree of local scientific autonomy surely retains its validity. Indeed, such a development may well contribute to the larger process by which nationhood itself is defined.

Given the universality of scientific knowledge claims, however, the idea of any nation establishing true scientific independence is surely chimerical. The scientific community of any nation, no matter how developed, does not operate in isolation but both contributes to and draws upon a common pool of scientific ideas and information. Thus what we should expect to find, with respect to nations of growing scientific strength, is not those nations literally becoming independent from a scientific point of view, but rather their emerging from a condition of scientific subservience to one where they are generally self-sufficient in scientific personnel and resources and where their leading practitioners enjoy some parity of standing and hence authority with those of scientifically established nations. Indeed, it could be argued that what is involved in thus achieving a condition of non-dependence in science is precisely the reverse of attaining a literal independence; a nation's leading practitioners are no longer marginal figures on the world stage, but become fully integrated into the international scientific community. Instead of authority being attributed automatically to some distant metropolitan centre, scientific participation and leadership become decentralized.

Paradoxically, therefore, one of the better indicators of a nation's scientists having achieved a state of non-dependence is the extent to which their work comes to be embedded within international rather than merely local networks. The nature and direction of communication across national boundaries is, however, crucial to our assessment here, since scientific colonialism is also characterized by strong links between one group of scientists and another. In the former case, it is an exchange among scientists of more or less equal standing, in the latter it is a matter of links between metropolitan patrons and scientists on the periphery who are being patronized.

The paradox involved here is heightened when it is recognized that embedding one's work within an international scientific network often implies not a gain but a loss of independence in regard to choice of research topic. Scientists working in a colonial environment frequently choose to investigate topics of immediate practical significance to their community,[5] whereas those who aspire to join the international network in their discipline are obliged to address problems of current concern within the network and these are usually set elsewhere, by already recognized researchers in their discipline. They thus face the danger that, in the very process of their work becoming less marginal to their peers on the international scientific stage, it becomes increasingly so at the national level. One of the best signs that formerly dependent scientists have established themselves in the international arena is when they themselves are able to introduce new topics into mainstream research in established disciplines; but they are less likely to be able to achieve this while working on locally orientated questions.

Despite what has been said above, some patterns are evident in the transition from a state of scientific dependency to one of non-dependency. When modern science first became established in parts of western Europe, international scientific exchanges were confined to relatively small groups in the nations involved. They collectively defined what constituted science, using patrons and publications to establish their authority, and they took advantage of expanding economic and political empires to export it. The process involved transplantation of ideas through time and space; and in the course of being transplanted, intellectual constructs and research techniques were inevitably inspected, expanded, muted, even challenged and discarded.

The multidirectional nature of scientific exchange was already evident in works of the late seventeenth and the eighteenth century. Prominent theories and intellectual systems depended on yet, at the same time dominated, the exploration of other parts of the world. Sir Isaac Newton's *Principia* included data acquired from Thomas Brattle in Massachusetts, while Carl Linnaeus's taxonomic system relied on a vast international data-gathering network with linkages through the various European empires to their colonized outposts. Usually in close alliance with politicians, merchants in international trade, and civil servants, European scientists took on the task of describing and defining the topography, plant and animal forms, and human inhabitants of the entire world in scientific terms. Commercial and diplomatic goals inevitably became intertwined with intellectual ones in the sometimes collaborative and often competitive efforts of naturalists and physical scientists from the different nations of Europe. Their labours established comprehensive although tantalizingly incomplete catalogues of plants, animals, rocks and minerals, along with maps of the earth and sky. The data they assembled contributed to new interpretations of geological time, topographical and climatic shifts, and world ecologies. Both the data and these new interpretations were disseminated internationally by means of published books and memoirs, and less formally through networks of correspondents.

As this newly accessible natural world was described, it also became a source of pride that science transcended boundaries between nations. When, during the French revolutionary wars, a British warship captured the great collection of Australian materials assembled by La Billardière, Sir Joseph Banks saw to it that this was released and sent on to Paris. Soon afterwards, despite continuing international tension, the Englishman Humphry Davy was welcomed in France and honoured by the Paris Academy of Sciences for his chemical discoveries, at the very time that his work was being hailed in Britain for demonstrating that chemistry was not, after all, a French science.[6]

Even today, when so much scientific work is directed towards war-like purposes and the interests of the state, the principle of freedom of scientific interchange is publicly asserted, for example through the International Council of Scientific Unions which in its statutes affirms 'the rights of scientists throughout the world to adhere to or to associate with international scientific activity without regard to race, religion, political philosophy, ethnic origin, citizenship, language or sex'.[7]

The fundamental interconnectedness of science is of central importance when historians compare settings for science across geographical space. The essays in this volume place the pervading issues of authority and hegemony within the context of a system in which communication, transportation and institutional networks are usually (although not inevitably) negotiated. Focussing for the most part on a single country, Australia, they set the atrophying of one set of extra-national linkages, in the transition from dependent 'colonial' to non-dependent 'national' science, against the growth of another set of international relationships in which Australian scientists play a less predictable and also generally less subservient role.

Australia offers a particularly appropriate location for analysis. The British initiated permanent settlements there at a time when both political and scientific imperialism were taking the institutional forms that would persist for a century and more. While scientifically trained émigrés from Germany and elsewhere contributed significantly to the scientific exploration of the continent, British imperial ties were always strong and British influences predominant in the structuring of local scientific culture. British lines of authority were established and maintained in subtle ways, by the transport of particular books or individuals, by the enthusiasm (or lack thereof) of imperial officials, by imperial networks of communication and publication, by colonial linkages with British institutions of education. Initially, of course, the settlers still thought of themselves as English, Irish, German or whatever, according to their country of origin. As a separate Australian political identity emerged, so did the goal of a distinctive intellectual identity, along with that of economic self-sufficiency and the search for an independent voice in diplomatic affairs.

What that identity might amount to in regard to science was, however, for the reasons already set out, much less clear than it seemed to be in relation to literature and the creative arts.[8] To be sure, beginning in the 1880s attempts were made to launch Australian scientific exploring expeditions to New Guinea and Antarctica – expeditions in which Australia would take on a new role as scientific centre in respect to the regions being explored as periphery. Yet in science there was nothing comparable to the strident nationalism that emerged in Australian literature and painting at this period. Indeed, it was precisely during the last decades of the nineteenth century that significant numbers of Australian scientists, most of them newly imported university professors, first began addressing internationally established research agendas instead of confining themselves more or less exclusively to locally defined practical problems or local taxonomy. Some taxonomic work, above all that of the botanist Ferdinand von Mueller, was fully integrated into the international networks and attained an authority of its own. Later, some of the new university appointees – for example the anatomist J.T. Wilson, the physiologist C.J. Martin and the anthropologist W. Baldwin Spencer – achieved international standing while exploiting research possibilities provided by Australian materials. In time, however, the best in science came for many Australians to be identified with internationally established agendas of 'pure' research, while investigations, necessary though they might be, that focussed on local problems, came to be thought of as second-rate. The trend had long-term consequences for the place of science in Australian society.

Just as the exercise of British imperial influence was sometimes direct and forceful and at other times more subtle, so the eventual realignment of scientific authority and lines of communication sometimes resulted in open antagonism while in most cases it occurred almost imperceptibly. The studies that follow are by historians who have been trained in

England, Australia, and the United States. Their accounts suggest that it is not enough to discuss colonial science in terms of isolation, inferiority, deference, or marginality. Such terms discount the capacities, initiatives, and ingenuity of scientists in the colonies. Early on, their collaborations with metropolitan colleagues rarely resulted in equal benefits to both sides. At least by the end of the nineteenth century, however, the vessel of colonial science was half full and rising.

This volume comprises a selection of the papers presented at a conference held at the University of Melbourne in May of 1988 on the theme, 'Nationalism and Internationalism in Science: Australia, America, and the World of Science'. Contributors take as their starting point existing discussions of the notion of colonialism in science, most notably those arising from an earlier Melbourne conference.[9] Some maintain the focus on the colonial condition itself. Others, however, shift our attention to the breaking up of the old lines of authority and colonial dependency by looking instead at Australia's expanding scientific links with a third power, itself of rapidly growing significance on the world's scientific stage, the United States of America. Here, Australian scientists, although not yet fully liberated from the ghosts of old Anglo-imperial connections, found themselves operating in an international arena where they had to fend for themselves in forging new intellectual relationships, in an environment that was unfamiliar to them and in which those with whom they had to deal had access to vastly greater resources. The papers in question illustrate how Australian science responded to these shifting patterns of intellectual authority within and beyond the Australian scientific community.

From the outset, David Wade Chambers reminds us, Australia's intimate human connections to England and to Europe counted enormously in all aspects of life. The large physical distances created some problems, of course, and Australia's remoteness had its effect with regard to political influence, the expensive transit of goods, and a sense of social and intellectual isolation that persisted throughout the colonial period and even after the acquisition of Commonwealth status in 1901. But what was the cost of distance in intellectual terms? Did distance in fact 'tyrannize' science, inhibiting its practice and constraining its recognition from earliest settlement even to the present day? Chambers critically examines the concept, popularized by the title of Geoffrey Blainey's book, *The Tyranny of Distance*, with particular reference to the diffusion and reception of scientific ideas. Australia's physical distance from Europe was far less inhibiting for science, he argues, than it has sometimes been taken to have been. Distance should be neither an explanation nor an excuse automatically used by historians of Australian science; rather, they must identify far more complicated explanations for the dependencies and distinctive initiatives that characterize Australian scientific work in the context of international science. Indeed, the colonial geographical context provided opportunities as well as limitations in physical as well as biological sciences, as other essays in this volume by Michael Osborne, Stewart Gillmor and Jan Sapp demonstrate. Distance, in other words, is quite different from what is historically significant, namely isolation. The latter may have many causes, and geographical distance is one of the least important of these.

Outsiders can feel remote from the intellectual discourse of their discipline, even when they are in geographical proximity to established centres of science. So David Knight argues in his discussion of scientific activity in England during the early nineteenth century. While physical distances, measured from London to the industrial North, to Scotland and to Ireland, were becoming less significant as a result of improvements in transport and communication, other powerful social and professional distinctions relating to class, education, and employment persisted. Social distance, in other words, could have at least as great an effect as geographical distance on scientific intercourse, and so too, later on, could the 'distance' between one science and another. Interlopers who tried to gain acceptance within the scientific circles of London experienced the hostility of the establishment. Even such exceptional individuals as Humphry Davy could not fully overcome this, although his appointments and memberships also indicate that patronage could make science a mechanism for social mobility.

Much depended, moreover, on the personality and intentions of those in a position to exercise authority. Some, like Richard Owen, whose positions at first the Hunterian Museum and later the British Museum gave him enormous prestige, were preoccupied with their own careers and provided little by way of return to the colonials who were part of their extended network. Elizabeth Dalton Newland investigates the relationship between Owen and the Australian colonist George Bennett in a case study that documents one of the more exploitative of scientific relationships and reinforces familiar notions of scientific imperialism. Her essay tracing the divergent careers of these two graduates of the Royal College of Surgeons suggests how difficult it was to maintain visibility and credibility for those who, like Bennett, chose to live in outposts such as Australia. The ambitious Owen, well situated in London, must have seemed to Bennett the obvious person to whom to send his first major discovery, the pearly nautilus. Significant as the sole living genus of Nautiloidea and related to fossil Ammonites, Bennett's specimen seemed a 'living fossil' and was of considerable value in contemporary geological debates. Owen's published description, under his own name, brought him rapid recognition and put him in an advantageous position to challenge the theories of his French rival Étienne Geoffroy Saint-Hilaire. Bennett, however, was relegated to the position of 'mere collector', despite his excellent descriptive work and his recognition of the importance of his specimen. Yet, curiously, the Australian naturalist did not turn against Owen, but continued to send his English colleague valuable specimens and an accurate (although not then accepted) account of the reproductive system of the platypus. The Owen-Bennett relationship confirms the reality of a conventional view that scientific colonists could be and often were exploited. Newland concludes that Bennett had little choice but to accept the subordinate status imposed on him by colonial domicile and paucity of European contacts.

Because of the extensive taxonomic and related theoretical work in the natural sciences in the eighteenth and nineteenth centuries, geographical settings with flora and fauna unfamiliar to Europeans received considerable attention. Local residents and visiting collectors in Australia, for example, gathered dried and living plants, skins and skeletons of animals, rock and mineral samples, and Aboriginal artifacts to be preserved, shipped, analysed and displayed in Western Europe and around the world. Much of the initial activity occurred informally through private correspondence and sales organized by field collectors and dealers. Gradually certain institutions and societies increased their role as overseers of scientific study and exchange. Official sponsorship and encouragement in the natural sciences became particularly important in areas of economic and political interest, as the case of geology reveals.

Robert A. Stafford demonstrates in his study of geological research in nineteenth-century Australia just how intimately this 'far frontier' was connected into the imperial system. The Geological Survey of Great Britain, particularly under the directorship of Sir Roderick Murchison, exercised a considerable degree of intellectual and practical influence over the practice of geologists operating at a distance from the London headquarters. Interpretation of geological data was dominated by British experts who not only dated fossils and evaluated economic minerals but also, on the basis of such data, classified strata into categories correlating with European precedents. The opinions of experts in London held sway, even in the face of important counter-ideas offered by less well connected colonials. The intellectual domination was reinforced by the appointment to Australian geological surveys of graduates from the Royal School of Mines and others who had worked for the British Survey. The practising geologists, though paid for by colonial taxes, were creatures of a patronage system conducted through appointed governors and others who made their reputation by being alert to economic opportunities presented by the newly explored land. Stafford maintains that these colonial scientists collaborated in maintaining a certain degree of dependence because there were evident personal benefits that flowed from this. Those with a 'garrison mentality' needed to do work so outstanding in conventional terms that it would earn them a position back home, while even those who stayed in Australia acquired positions of prominence based at least in part on the recommendations they had brought with them. This was the imperial system at its most efficient.

Michael A. Osborne looks at imperial relationships from a different viewpoint. He points out that while some attention has been paid to the practice of science within the empires of Germany, France, and Britain, little is known about the ways in which some scientific efforts actually transcended these systems. Focussing upon the French and British empires and the 'glamour settlements' of Algeria and Australia, he analyses the membership, organizational structure and policies of leading acclimatization societies and explores their fundamental scientific assumptions about the nature and distribution of species. The Société Zoologique d'Acclimatation, founded in Paris in 1854, exercised considerable authority over France's other metropolitan societies with an interest in acclimatization, as well as over branch groups in various French colonial settlements. By contrast, the London Society for the Acclimatization of Animals, Birds, Fishes, Insects, and Vegetables was largely ineffective. As one result, independent colonial acclimatization societies in Melbourne and Sydney were far more active and successful. Not surprisingly, therefore, the Parisians tended to solicit plants and animals for Algeria from Australia directly, the extent of their activities being most dramatically illustrated by the million and more eucalyptus trees planted in North Africa.

This co-operation across imperial and cultural boundaries did not however come from or even facilitate shared interpretations about the nature of species, a topic presumably fundamental to acclimatization practice. The leading Australian acclimatizers, particularly British-born Frederick McCoy and German-born Ferdinand Mueller, held to an anti-transformist position and the concept of regional centres of creation of organic forms. This was in contrast to ideas about the flexibility of organic forms held by their French counterparts, Charles-Victor Naudin and Isidore Geoffroy Saint-Hilaire. Thus, according to Osborne, while acclimatizers bridged political chasms in the common pursuit of redistributing living species across geographical zones, the theoretical interchange was negligible.

Much easier was the exchange of pressed plants, dried skins, skeletons, and fossils that added data to taxonomic systems and permitted independent theorizing about geological time and the relatedness of species. If colonial naturalists were on the intellectual and geographical fringes of taxonomic research, they were essential components in a system the connections within which reached vertically from colonizer to colonized rather than horizontally from one colony or former colony to another. Private European collections were transformed into co-operative or public museums and botanic gardens during the nineteenth century as the influx of material grew beyond the means of even wealthy naturalists. The government-sponsored Royal Botanic Gardens at Kew and, to a lesser extent, the British Museum (Natural History), initiated considerable collecting activities throughout the British colonies, abetted by such private groups as the Linnean Society.

Sally Gregory Kohlstedt's study begins at the other end of this story, tracing the development of natural history museums in America and Australia. Their initial functions included serving as a conduit for the export of specimens and information, but quite quickly they established more self-directed goals. Thereafter, Australian and American museum staff held parallel aspirations for their museums in their desire both to be useful to their local community and to gain international recognition; but the two countries provided disproportionate means to pursue these goals. When it came to joint investigations, American institutions gained an advantage because private patronage figured so largely in their developmental programme, whereas Australian museums remained almost totally dependent on often parsimonious colonial governments. In the twentieth century, sustained contact between museum scientists and their institutions in the two former British colonies helped pull Australians away from their British orientation, even as they struggled to establish their own authority over locally derived materials.

How bonds of empire were maintained in the field of science is the theme of R.W. Home's paper, which traces the election of Australians and, by way of comparison, other colonials to Britain's premier scientific institution, the Royal Society of London, during the heyday of the British empire and subsequently. He points out that scientists throughout the empire (and, later, the Commonwealth) have always been viewed as members of the British scientific 'family' and thus eligible for election as ordinary members of the Society. Still, during the nineteenth century Australians were quite distant cousins, and their uneven success in the Society's elections suggests that they were under more disadvantage than, for example, expatriate scientists resident in India, where a large civil service and frequent exchanges of personnel meant that potential members could more easily assemble sponsors familiar with their work to nominate them. The situation improved after 1880 in ways that reflect the growing strength of science in the Australian universities. Evidence gleaned from the statements of nomination for new members indicates the importance of personal contacts between colonial scientists and those 'at home'. A wave of new elections from the colonies around the turn of the century made the Royal Society an institution of genuinely imperial scope. More recently the Society, by continuing to offer the possibility to scientists from Britain's former colonial territories of being elected to its Fellowship, has, Home argues, helped Britain retain a degree of hegemony over the science of those territories, even after the breaking up of the empire. Only in very recent times have the first signs appeared of these traditional bonds beginning to crumble.

While certain affinities with Britain thus remained very strong, including not just the ongoing prestige of Royal Society membership but also links deriving from the fact that large numbers of Australians continued to undertake their advanced training in England, by the Second World War the United States had also attained considerable influence within the Australian scientific community. The trend was already clear as Americans extended their biological field research efforts into the antipodes. Over time, American-Australian museum connections grew stronger, as Kohlstedt's essay points out, bolstered by the financial resources of American universities and foundations. By the 1930s various collaborative efforts, some sponsored by Carnegie Corporation endowments, were evident in both research and exhibition activities, suggesting that to some extent the orientation of the Australian curators had turned toward colleagues in North America. In the final four chapters the growing American involvement with various other aspects of Australian science is apparent. It is equally apparent, however, that these new linkages were of a different character than the longstanding Anglo-Australian ones.

Certain sciences depend on systems and data gathering that extend beyond conventional geographical and even intellectual arrangements. The gathering of specimens from around the world through ad hoc and informal mechanisms contributed to the taxonomic enterprise of the eighteenth century, and the results underpinned important evolutionary theories in both geology and biology in the next century. In the twentieth century, research in ionospheric physics has depended on a degree of international co-operation and quite early elicited participation from Australians, many of whom were involved in the establishment of Australia's telecommunications systems. In time, this became one of the earliest fields in which traditional Australian scientific linkages began to wither away.

Radio research was the only field of research not linked to agriculture to be sponsored by Australia's new Council for Scientific and Industrial Research, founded in 1926. C. Stewart Gillmor discusses the important work conducted by the Australians in their own research facilities even as his extensive survey data document their close connection with British colleagues by virtue of shared educational and professional affiliations. These traditional scientific networks were complemented rather than superseded by increasing connections to Americans through the Carnegie Institution of Washington – which operated a geomagnetic observatory at Watheroo, Western Australia – and, later, the United States National Bureau of Standards. Gillmor's analysis shows how, through these growing multilateral linkages, along with their participation in the Allied collaboration aimed at enhancing war-time radio propagation services, Australians came to be accepted in the post-war world as fully-fledged members in their own right of the international community of radio scientists.

The special geographical advantages enjoyed by Australian ionospheric physicists as a result of their location on the 'underside' of the globe play a significant role in Gillmor's analysis. In another area of geophysics, too, Australians have often been held to have had the benefit of location from which to make specific and unique observations. A commonplace among geologists, H.E. Le Grand explains, has been that scientists in the southern hemisphere accepted the theory of continental drift much earlier than those in Britain and North America because, as one wag recorded, 'most of Earth's biggest scars, sutures and dimples etc. were more visible on her bottom'. Le Grand challenges the historical basis of this claim, pointing out that it was neither the case that Drift was unanimously opposed by northern hemisphere geoscientists, nor that it was widely favoured by their southern hemisphere counterparts. Le Grand traces the emergence of Australian support for Drift in the 1950s and 1960s, carefully placing it within the context of contemporary geological work being done by others, both in Australia and elsewhere. He shows that Australians were not disproportionately represented among those engaged in such activities, but that their work was quite visible, and concludes that by this period and in this field, Australian workers, like their ionospheric physicist cousins, are best viewed simply as part of a rapidly developing international network of research activity. The traditional dichotomies of centre-periphery, metropolis-colony and so on do not apply. Le Grand acknowledges that, at the time, Australia's rocks were 'as intriguing as its bizarre flora and fauna had been in the nineteenth century', and that some Australian (and South African) earth scientists did do work that was somewhat unorthodox. He concludes, however, that the commonplace view of the history of Drift is nevertheless a myth, one that helps preserve a mythical view about scientific progress more generally: 'It provides a satisfying account of why the majority opinion in the international community from the 1920s through the mid-1960s was, as judged from today, wrong; for it shows it to have been wrong for good reasons'.

The migration of individuals has played a critical role throughout the history of Australian science, especially when those individuals have found themselves for whatever reason committed to an extended stay in Australia. The British and German naturalists who moved to Australia in the nineteenth century established research agendas as well as institutional bases that served not only their career goals but also local enthusiasts and new immigrants. As is most clearly illustrated in the migration of refugee scientists from central Europe to North America in the 1930s, though the factors leading an individual scientist to move to another country may be far removed from his or her science, they may nevertheless have a profound impact on the science in question. In twentieth-century Australia, immigration of key individuals has likewise played an important role in the development of the universities, the CSIRO and various other scientific institutions into well-established centres of research operating in ways parallel to if not precisely on a par with similar agencies in North America and Europe.

Jan Sapp in his chapter describes how personal, institutional and political problems led the eminent cytogeneticist Michael White to move first from England to the United States and later, as a result of McCarthyist pressure during the 1950s, to Australia where very little genetics had previously been done. Sapp shows how White subsequently helped establish genetics in Australia and in particular how, with the standing he already enjoyed in this field, he was able to integrate his new department at the University of Melbourne into the international genetics research network. At the same time, Sapp suggests, White's physical remoteness in Australia from the dominant research groups in the US may actually have facilitated his challenging the established 'evolutionary synthesis' during the 1960s and 1970s with his unorthodox but influential ideas on modes of speciation. Indeed, White's moving to Australia may have advanced his work in an even more concrete way: 'the origin of White's theory of stasipatric speciation cannot be understood', writes Sapp, 'without reference to the Australian group of grasshoppers that helped to shape it'.

The final chapter, by Stephen J. Pyne, contrasts the very different approaches developed in Australia and the United States during the twentieth century to forest management in relation to fire. In Australia, both research and fire service practice concentrated on the use of controlled burning; in the US, the idea of controlled burning came to be anathema. US Forest Service policy came to be framed instead around the notion of fire suppression, while research mimicked Big Science, 'merg(ing) field trials on large fires with "analytical" solutions encoded into fire behaviour models'. Pyne argues that these different approaches were intimately linked with very different patterns of population distribution and land use, so that when American forest science joined the general American cultural invasion of post-Second World War Australia, it was firmly resisted by leading Australian foresters. Pyne sees in this not merely a deeply held Australian scepticism about the transferability of American methods to Australian conditions, but also 'a stubborn if inchoate Australian nationalism' that involved 'a redefinition of Australian identity in terms of the bush'. In Pyne's account, in both countries scientific research has been inextricably bound up with social structures and cultural values.

Ironically, in recent years Australian foresters have been driven by changing demography and attitudes towards the natural environment to move somewhat towards the American approach, precisely as changing American attitudes have driven foresters there to incline more towards the Australian approach. Here, then, we have once again an account in which ideas such as metropolis and province, or centre and periphery, no longer have any place. Australian foresters were well aware how far their ideas differed from those of their American colleagues. They were not however induced thereby to change them, but stood firm concerning the validity in the Australian context of their own approach. 'When Australians and Americans talk fire', Pyne concludes, 'almost always they speak past one another'.

The level of scientific activity in Australia has risen steadily since the Second World War, measured by a very substantial growth in numbers of scientists as well as increasing visibility and influence in those countries with which Australia has most regular contact. England was and to some extent remains a key reference point, but American connections have become particularly strong in areas like genetics and astronomy. As the links with Britain have become more diffuse, Australian scientists have become increasingly integrated into the international networks of their disciplines, helping to set agendas as well as to follow them. The converse of dependence has proved to be not independence but interdependence, not national science but international science. What has changed in the course of the twentieth century, so far as Australian science is concerned, is the distribution of authority within particular scientific fields. In different fields at different times, Australian scientists have come to claim, and to be granted by other workers in those fields, a measure of authority of their own.

R.W.H. and S.G.K.

NOTES

[1] An entire recent issue of *Isis* dealt with the topic of scientific setting, beginning with Jeffrey R. Sturchio's 'Editorial: Artifact and Experiments', *Isis*, 78(1988), 369–72. Issues of authority and control were explicitly raised by Edward A. Shils's arguments in 'Center and Periphery', in *Logic of Personal Knowledge* (Glencoe, Illinois, 1961). Historians of science have sometimes extended and more recently also reconsidered the dynamics of scientific centres and the hinterlands. On Europe see Mary Jo Nye, *Science in the Provinces: Scientific communities and Provincial Leadership in France, 1860–1930* (Berkeley, 1986). On Australia, R.W. Home, ed., *Australian Science in the Making* (Sydney, 1988). On colonial relations see Lewis

Pyenson, *Cultural Imperialism and the Exact Sciences: German Expansion Overseas, 1990–1930* (New York, 1985); Lucile Brockway, *Science and Colonial Expansion: The Role of the British Royal Botanical Garden*, (New York, 1979); and Susan Sheets-Pyenson, *Cathedrals of Science: The Development of Colonial Natural History Museums during the Late Nineteenth Century* (Kingston/Montreal, 1988).

[2] John C. Greene, *American Science in the Age of Jefferson* (Ames, Iowa, 1984), esp. pp. 5–12; and Nathan Reingold, 'Reflections on 200 Years of American Science', in Reingold, ed., *The Sciences in the American Context: New Perspectives* (Washington, D.C., 1979).

[3] George Basalla, 'The Spread of Western Science,' *Science*, 156 (1967), 611–22.

[4] Roy MacLeod, 'On Visiting the "Moving Metropolis": Reflections on the Architecture of Imperial Science', *Historical Records of Australian Science*, 5(3) (1982), 1–16; reprinted in Nathan Reingold and Marc Rothenberg, eds., *Scientific Colonialism: A Cross-Cultural Comparison* (Washington, D.C., 1987), pp. 217–32.

[5] Ian Inkster and Jan Todd, 'Support for the Scientific Enterprise, 1850–1900', pp. 102–32 in R.W. Home, ed., *Australian Science in the Making* (Sydney, 1988); Roy MacLeod, 'The "Practical Man": Myth and Metaphor in Anglo-Australian Science', *Australian Cultural History*, 8 (1989), 24–49.

[6] Gavin de Beer, *The Sciences were Never at War* (London, 1960), chaps. 5, 13; David Knight, 'Revolutions in Science: Chemistry and the Romantic Reaction to Science', in William R. Shea, ed., *Revolutions in Science: Their Meaning and Relevance* (New York, 1988), pp. 49–69.

[7] [International Council of Scientific Unions], *Advice to Organizers of International Scientific Meetings* (1987), p. 1.

[8] Geoffrey Serle, *From Deserts the Prophets Come: The Creative Spirit in Australia, 1788–1972* (Melbourne, 1973).

[9] Reingold and Rothenberg, op. cit. (n.4).

DAVID WADE CHAMBERS

DOES DISTANCE TYRANNIZE SCIENCE?*

MYTHS OF SCIENTIFIC ISOLATION

Twenty-seven years ago, in 1961, the art historian Bernard Smith presented a lecture at the University of Queensland which he entitled 'The Myth of Isolation'.[1] In that address he argued that the idea of cultural isolation has become an obsession for many Australian commentators, whose 'wailing and beating of the breast' betrays a deep sense of inferiority, an early reference to what Australians have come to call the 'cultural cringe'. Can there be any doubt that Smith's critique, although primarily concerned with the world of art, has broader application? Without perhaps beating their breasts about it, Australians have, I think, for too long accepted the myth of distance-imposed isolation as a general explanation for supposed weaknesses in their own scientific tradition.

Whatever influence Smith's lecture may have wielded in Australian art circles,[2] his phrase 'the myth of isolation' was never to capture the popular imagination as did the phrase 'the tyranny of distance' when it was introduced five years later by historian Geoffrey Blainey.[3] Blainey's cringing metaphor hit such a responsive chord with European Australians that it soon passed into the language and into popular culture. To this day the metaphor is offered up as both explanation and excuse. It remains, in Rod Home's words, 'a recurring theme in recent writing on Australian history'.[4] Why are Australians so ready to believe that their potential intellectual achievement is constrained by their distance in miles from the rest of the world, most particularly Europe, as well as by their distance in miles from each other?

Bernard Smith acknowledged, as we must here acknowledge, that there is truth at the heart of any myth. Doing science 'beyond the black stump' raises problems not encountered when doing science in, say, London. But many of these problems have been exaggerated, especially with reference to the last one hundred years. The image of a remote settlement in a hostile environment, isolated, in one writer's dramatic phrase, both 'without and within'[5] has haunted European Australians from the

R.W. Home and S.G. Kohlstedt (eds.), International Science and National Scientific Identity, 19–38.

beginning of European settlement, and for many the ghost has yet to be exorcised. While distance and geographical isolation may have played a major role in the generation of an Australian world-view, these factors have not by themselves greatly impeded the establishment of European science in any colonial setting and certainly not in Australia. In this country geographical constraints were at their most significant in the period before scientific institutions were established, and have exerted only marginal influence ever since. I shall also argue that some problems of isolation, supposedly related to distance, have more to do with social, cultural and economic factors than simply with accumulated mileage from some perceived centre.[6]

I want first to consider the very general question of cultural isolation, the hypothesis that Bernard Smith was determined to refute. Consider, for example, the following dramatic account of the supposedly dire effects of geographical isolation on science: 'For Australia, more than almost any other country, distance seems a central and common factor in its history ... Australia is as far from the European origins of the great majority of its settlers as it is possible to be; a separation not only from family but from the history, culture, ideas and institutions of their homeland'.[7] The precise mechanism by which such factors exert their influence is rarely suggested. Ideas and institutions will be considered in detail below, but what are the real effects on science of cultural isolation? And are Australians truly isolated from family, history and culture?

Although the rise of modern transport and communication proceeded rapidly through the various stages of colonial development, many writers still seem to suggest that the country suffered from cultural isolation throughout its history. Notice that despite the fact that the subject of the paper quoted above is confined to 'the earliest years of the twentieth century', its comments on isolation are couched in the present tense and the author asserts that 'distance is a central and common factor in [Australian] history'. Such comments seem particularly inappropriate in the present day since the majority of Australians have never been to Europe, and increasing numbers are not even of European origin.

While the cultural *perception* of isolation and distance may some day be shown to be a central factor in Australian intellectual history, I find it difficult to understand how, considered by itself, the diffuse notion of cultural isolation (one might almost call it a homesickness factor) is thought to inhibit the doing of science. In the eighteenth and nineteenth centuries, the institutional structure of science was based in Europe. Thus, insofar as they attempted to do science, the settlers' cultural and intellectual origins in Europe were an advantage rather than a disadvantage. Far from 'home' they may have been, but they were nevertheless 'more British than the British'. To paraphrase Bernard Smith's comments on painters, 'our colonial [scientists], despite the time and distance from Europe, were certainly not isolated from the European tradition ... [They] were European [scientists] in Australia ... In order to belong to a tradition it is not necessary to have one's nose rubbed in it three times a day.'[8]

Barry Butcher has emphasized this point in a recent paper in which he describes Australian science as constructed 'by Europeans, within a European tradition and with constraints imposed by that tradition'.[9] George Seddon has briefly explored, in a series of highly concrete examples, how the fundamentally European perspectives of colonial Australian scientists affected their theoretical approach, providing a fully imported framework of institutions, ideas, analogies and classification systems. In a sense Seddon turns the distance metaphor on its head (just as he says he usually reverses globes when he passes them), suggesting that science was retarded not because Europe was so 'far' but rather because it was so 'near'.[10] The significance of the European-ness of colonial scientists will be further considered below, but here I simply note that the scientists' cultural orientation and tradition, rather than isolating them, actually worked to increase their social access to the hallowed halls of European science.[11] It may also be pointed out that their totally Eurocentric orientation left them with no adequate way of understanding and valuing indigenous scientific knowledge, a failure that is felt to this day.[12]

But let us turn to the straightforwardly practical side of the problem of distance. Can we find in Blainey's *Tyranny of Distance* an account of particular social and intellectual constraints imposed by Australia's antipodean separation from Europe? If one first comes to the book with such expectations, as I did, one may be surprised to discover no serious or sustained discussion of distance, as such, or of the myriad of expected effects of geographical isolation on Australian history. From beginning to end the book is a popular history of Australian transport and little else. I do not wish to criticise this fact about it. Rather, I simply suggest that the book is not what it is sometimes taken to be: it is not an exploration of the influence of Australia's geographical position on the formation of the national character or on the evolution of its political, economic and educational institutions. It is not, nor does it pretend to be, an analysis of the role of geographical distance in Australian cultural and intellectual history. There are some, though not many, references to the role of distance in its economic history.

Consider, for example, the chapter entitled 'Isolation', which discusses such matters as navigation in the roaring forties, the provisioning of the young colony, and commerce with Asia and the Pacific Islands. This is a limited and straightforward account of the technologies of travel and commerce, that gives the reader only the most occasional and barest traces of Blainey's ideas on how these geographical and technological realities actually influenced the distinctive evolution of Australian institutions or the development of a characteristic world view.

It is not entirely unfair to raise these more general questions about *The Tyranny of Distance*. After all, Blainey invites such comparison when he suggests in the preface that 'in understanding Australia's history, the idea of distance may be as revealing as say Frederick Jackson Turner's "frontier theory" is in probing the history of the United States'.[13] Indeed it may be, but not in Blainey's hands. Blainey's *Tyranny*, which followed the frontier thesis by three-quarters of a century, shows neither Turner's breadth nor depth. Turner attempts to explain a number of apparently distinctive aspects of America's social and cultural development. He looks at the influence of the 'frontier' on politics, legislation, indeed on democracy itself. He examines the relationship of the people to the land and of nature to culture, and the influence of the environmental encounter. He proposed the idea that the 'frontier' acted as economic and psychological safety valve. He may be said to have explored in American history the dialectical relationship of individualism and community, of materialism and idealism, of innovation and conformity, of nationalism and regionalism, of pacifism and militarism, of the urban and the rural, and of the primitive and the civilised.[14] None of these dimensions of historical explanation did Blainey attempt in *The Tyranny of Distance*.

In truth, Blainey's title leads the reader to expect a series of questions that are not in fact addressed. For instance, has the distance from the 'motherland' influenced Australians' use of language, their hopes and aspirations, their fears, or their perceptions of themselves? Has distance made them more creative or less, more individualistic or less, more conservative or less, more tolerant of other races or less? Are Australians, because of distance, more self-reliant, more ingenious, more dubious of foreign expertise, less receptive of imported ideas? Does Australia's distance from Europe, and its consequent perception of isolation, tell us anything about the differential political, economic and intellectual development of the relatively distant colonies, such as Australia and New Zealand, as opposed to the relatively near, for example Canada and the USA? It seems to me that if Blainey's book had sparked an examination of some of these issues, we should now be clearer about certain questions in the history of Australian science and technology.

Blainey wrote not of the *tyranny* of distance (or isolation and its effects), but of the *taming* of distance through 'its enemy, efficient transport'.[15] He makes this abundantly clear in his preface, saying that the last half of the book, which carries the name 'The Taming of Distance', incorporates his original intention 'to write on the coming of mechanical transport to Australia – steamships, railways, aircraft, automobiles – with some attention to their effects'.[16] With regard to this limited objective, he admirably succeeds. And the story he tells of technology triumphing over distance is a story that may be told for every country on Earth, not excepting the little island of Britain.

Before the book was finished, Blainey says, he realised the need to explain the effect on Australia of isolation. In this he is perhaps less successful: the first half of his book, which he called 'Destiny and Distance', is little more than the history of the sailing ship and the bullock in an earlier period of Australia's transport history. So we are left with a paradox. Twenty-two years ago Blainey published a lively account of how distance, during the course of the nineteenth century, was brought to heel. Yet, ever after, because of his memorable title, he has been credited (and not only in popular circles) with precisely the opposite achievement: showing how Australia always and ever remains tyrannized by distance.[17]

ISOLATION FROM IDEAS AND INSTITUTIONS

If we accept Blainey's account of the historical time frame for the taming of distance, then we must accept a significant fact about the history of Australian science: almost from the time that Australia had any science at all (i.e. from the time the first scientific institutions came into being), isolation due to distance alone was already on the downturn. Home and others have described how the advent and development of rail, steamer and telegraph improved scientific communication within Australia.[18] Internal distances in Australia were no greater than in the USA, though of course the population was much smaller. The problem in both places was the lack of a scientific capital. Talent was spread among too many centres to provide proper focus and maximum interaction. This structural problem resulted not from topographical necessity but from political and economic choices.

Scientific and intellectual isolation occurs in every country, however, and for many different reasons. In many places language has been a far greater isolator of scientific thinkers than distance could ever be. Furthermore, we know from studies of seventeenth-century scientists that social class isolated many people from effective participation in science.[19] Or consider the isolation that comes simply from being an 'outsider' (often so categorized for personal or social reasons). Outsiders may appear to be wonderfully placed in a Boston-Washington or London-Oxbridge institution. Then too, disciplinary boundaries isolate people often in the most direct and immediate ways. No one is likely to quarrel with the notion that isolation, whatever its cause, may have significant effects on the history of science. But we are entitled to question the extent to which distance in kilometres – between the scientific centre and the scientific periphery – is likely to be, or ever has been, a major factor in creating that isolation.

If we are to discuss the distance between centre and periphery, we must define what we mean by scientific centre. During much of the nineteenth century, one might suggest, Europe (especially France, Germany and the UK) was the centre and everywhere else the periphery. But of course Europe itself, indeed London itself, contained peripheral institutions, and on the other hand the peripheral nations, increasingly, supported scientific centres. These complications alone make such analysis both difficult and suspect. I intend to offer a more precise definition towards the end of this chapter, where I set out a rather different way of looking at these problems. For present purposes I mean by 'the scientific centre' to denote that nineteenth-century association of people and institutions (societies, academies, universities, journals and so on) existing in a small number of European cities. To use the words of the editors of the *Scientific Colonialism* volume, science is 'polycentric'.[20]

What were the problems of doing science outside that centre? What sort of access could one gain to the centre from without? And what was the role of distance in aggravating those problems and in hindering that access?

One of the first problems encountered when working outside the centre, whether this century or last, is the attitude, dare I say contempt, of some of those firmly ensconced in the central institutions. I suppose this has been so since the city mouse first encountered his country cousin. We may recall the story told by Susan Faye Cannon of Henry Brougham's 1803 *Edinburgh Review* article assessing an early volume of the *Transactions of the American Philosophical Society*.[21] Brougham began with the comment that the volume would not 'repay the labors of him who may be induced to wade through it'. He ended his review with the even more scathing comment on 'a curious and unaccountable fact – the scarcity of all but mercantile and agricultural talents in the New World'. Within the review he had belittled the thoroughly competent work of Andrew Ellicott, who had reported on a survey expedition, that Cannon characterizes as 'the best and probably the most ambitious enterprise involving the accurate use of instruments in North America during that period'. Two of the best-performing instruments used by Ellicott, incidentally, were zenith sector telescopes by the American David Rittenhouse.[22] We can thank Cannon and many others for showing us that American colonial science was not so backward and unsophisticated as Europeans, and indeed many Americans, presumed it to be.[23]

Or consider the case of European attitudes to Alexander von Humboldt's comments on the remarkably advanced research and teaching at the Mexico City School of Mines, in which Humboldt had worked for nearly a year (1803–1804). His high praise of the Mexican institution was simply not accepted by the European reviewers of his *Political Essay on New Spain*.[24] Even Humboldt's English translator John Black says in his preface to the *Political Essay* that Humboldt's favourable accounts of both scientists and scientific institutions in Mexico were nothing more than a repaying of the hospitality he had received there.[25]

But for evidence of central disdain for the periphery, one need not to go so far as the USA or Mexico, much less Australia. Consider the sharp words sent to London from Ireland by G. F. FitzGerald relating to the control of the Royal College of Science of Ireland: 'I know the contempt London has of all provincial places and how they think they can arrange things for us so much better than we can for ourselves that I don't believe they will give up one jot or tittle of power'.[26] Australian examples of European contempt for colonial scientists are not hard to find, and indeed the well known phenomenon of the 'cultural cringe' marks Australian complicity in this assessment.[27]

The sad truth is that historians of colonial science have often, sometimes unavoidably, encouraged this sort of 'cringing' assessment by describing colonial scientists almost exclusively against the backdrop of *European* science, rather than by painting them in the actual scientific landscape in which they lived. Thus, by definition, colonial scientists become 'second-rate'. Donald Fleming, from whose writings we have all immensely gained, often rose above this tendency, yet never fell lower than in the following words: 'Colonialism in science owed much of its tenacity to academics of the second rank, Englishmen who did not succeed in rebounding from exile, native colonials who had no choice but to return home after their studies abroad because nobody tried to detain them'. This is certainly, as Joseph Needham said in a slightly different context, 'very near the edge'.[28] It vastly oversimplifies the motives, impugns the intellectual stature, and undervalues the contribution of colonial scientists. Whenever historians have found yardsticks of scientific merit that measure more than estimated rank on some sort of European 'Richter scale' of scientific achievement, they have thereby provided a more balanced picture of colonial scientific work.[29]

Whatever may be said of European disparagement of colonial accomplishment, it is certainly possible to cite examples of European collaboration and support for their far-flung colleagues. Humboldt, interestingly, shows traces of both tendencies, at times praising lavishly, at times undervaluing the work of particular scientists.[30] The point is that those associated with the centre are more likely to be taken seriously than those who are not. Colonials, and indeed all provincials, must prove their worth. In any case, I have found no evidence to suggest that disdain as a cause of professional isolation is now, or ever has been, greatly compounded by distance.

A second way in which geographical isolation has been said to play a role has been through the delayed transmission of ideas from centre to periphery. Undoubtedly invisible colleges and professional networks (to use modern terminology) have played an important role in the diffusion of ideas. Exclusion from these networks in some measure entails exclusion from the latest ideas, at least from the *very* latest ideas. If James Watson's book tells us nothing else about science, it shows that, in some circumstances, spectacular scientific success may stem from knowledge of the very latest ideas and techniques learned from chance conversations, comments overheard in seminars, the circulation of unpublished reports, contact with the son of your chief competitor, fortuitous office sharing, and in general rubbing shoulders with intellects that are focussed on the same problem.[31] Needless to say, one need not go all the way to Australia to miss out on ideas transmitted in this fashion.

More commonly, and I presume much more importantly, ideas are transmitted through publication in books and journals and through scientific correspondence. Fortunately, these modes of communication are anchored in artifacts that are frequently preserved. We can consult booksellers' lists, catalogues of private and public libraries, published reviews of books, comments in private correspondence, educational textbooks, peripheral editions of centrally produced books, indeed all the rich veins that may be mined by the social and intellectual historian. In many former colonies this work has only just begun.[32] Nevertheless, a clear picture has begun to emerge of a much smaller transmission gap than had been traditionally assumed; and that gap is usually due more to cultural obstacles than to the technology of communication across distances.[33]

It is not uncommon for major and minor scientific works to be republished in peripheral nations within a few years of their first appearance.[34] But why is one European book republished and not another? In the few colonial countries with which I am most familiar, the answer has much less to do with isolation or distance than with conscious choice, based on cultural and intellectual preference and on sensible, perhaps tacit, perceptions of the political and economic interests of particular social groups.[35] Thus a directory of such publications for any country, with all its interesting inclusions and exclusions, would provide evidence not of transmission delay or ignorant isolation, but rather of a country's intellectual profile and of the cultural behaviour of social groups within that society. The same sort of analysis, of course, applies to the library holdings, the citations and reviews, the school curricula, the scientific correspondence, and ultimately the scientific ideas themselves.

One way of tackling the question of the diffusion of ideas is to examine the array of 'reception studies' that have appeared in the last fifteen years.[36] To look, for example, at the comparative reception of evolutionary ideas is to give up forever the classical diffusion model – we might call it 'the parcel theory' – which says that (1) a new idea appears which (2) is communicated through certain professional channels (3) over time and distance (4) destined for particular members of the professional community. Significant problems exist at every stage of this simplistic linear model,[37] and thus each of the four stages deserves separate treatment.

(1) How do we define the unit idea?[38] Evolution is perhaps the most

complex set of ideas ever to occur to the human race, not even excepting the ideas of physics. The most elementary discussion of the notion requires some familiarity with animal and plant taxonomy, with comparative anatomy, with paleontology and geological processes, with embryology and genetics. Before we talk transformation of species we must talk species, we must talk continuity and change, development and progress, competition and co-operation. Many of the richest strands of human thought are bound up in the idea. Of course, what I am saying is not new; it is the very bread and butter of many contemporary scholars. My point here is to emphasize that, in explaining the diffusion of Darwinian evolution, our task cannot be simply to date the first appearance and subsequent scientific acceptance of the *Origin of Species* in a particular country. Such an approach would leave us forever mystified by its slow diffusion to France and by its virtual non-appearance in the Islamic world until at least the 'eighties.

(2) The channels by which scientific ideas get communicated are well known (and were referred to above), but several considerations here must not be lost sight of. Even when ideas are being transmitted to the far periphery, the channels are not one-way streets. The channels themselves are multi-branched and complex. The channels themselves may influence or obscure, speed or slow the message. Most importantly, the channels are often under the control of the scientific centre rather than the periphery.[39] In the case of the *Origin*, there were of course journal reviews of the book, the book itself, other books, correspondence among scientists and intellectuals, and institutions such as museums and universities.[40]

To illustrate how these channels of authoritative communication may be extremely complex, consider one case of the crossing of boundaries of language and culture. Mexico, in this sense, may have been much *further* away from the scientific centre than Australia. Mexico heard about Darwinian ideas mainly by way of France.[41] Through most of the 1860s in Mexico not only was there intense civil disturbance, disruptive of all intellectual discourse, but a direct French intervention by Napoleon had enthroned a European princeling. During this intervention, the French Ministry of Public Instruction established an influential scientific commission that had a lasting influence on Mexican science.[42] In any case, Mexico had long before learned that the most direct line to the latest European ideas was through Paris, not Madrid.[43]

Secondly, the channels may give confusing or contradictory messages. This is in part because scientific messages are communicated with varying degrees of authoritative support and dissent, subtleties that may be lost or confused in the transmission. This factor has much less to do with distance between centre and periphery than with the particular networks or channels on which a particular scientist may depend. An excellent example of this phenomenon, extremely revealing for centre-periphery relations, involves the reception of Darwinism in Australia. Those whom one might call the 'institutional' scientists, men like McCoy, Halford and Macleay, who owed their positions to support from the European centre, were in fact creationist followers of Owen and thus opposed Darwin. Clearly such scientists as these had access to institutional channels unavailable to the 'unconnected' scientist. However, outside the institutional framework, other working scientists, like the anthropologist Howitt and the botanist Fitzgerald, quickly accepted Darwin's ideas as their European correspondence indicates. In the 1870s their sophisticated scientific work, informed by their Darwinian stance, demonstrates that they were professionals of some stature, not simply collectors in the supposedly colonial tradition.[44] Whatever else this episode may show, it clearly demonstrates that connections with the scientific centre are not simply constituted of the parcelling and transmission of information from centre to periphery.

(3) A survey of the various countries examined in Glick's Darwinian anthology shows that distance had little or nothing to do with the time taken for the full transmission of Darwin's ideas. The USA and Russia were particularly quick off the mark whereas France, though aware of Darwin, was slow to give his work significant discussion. Spain was even slower, as was Mexico, for reasons cited above. Significantly, even in London the *Proceedings* and the *Philosophical Transactions* of the Royal Society had no full discussion of Darwin's theory before 1870. Darwin's name is scarcely mentioned.[45] This means we must hesitate before offering apologies for the non-appearance of Darwin in the colonial scientific literature before the 'seventies. By the 'seventies, in countries such as Australia, Spain and Mexico, discussion of Darwin was well under way.

In Australia, the question of the delayed transmission of scientific ideas due to distance has generally been assumed by historians to be significant. Home, discussing Bragg, points out that by the late nineteenth century the delay in the mails had been cut from an 'impossible' fourteen weeks to an 'inconvenient' six weeks.[46] I find this sort of calculation unconvincing.[47] In highly unusual cases, I suppose, even a day or two might make a difference. One colleague of mine suggests that what matters is not the time between London and Melbourne but rather the time between ships. Writing of the last half of the nineteenth century, Graeme Davison says, 'Melbourne was a distant but *constant* satellite of Britain and the United States'.[48] It seems to me that it is the constancy of the information flow and the regularity of exchange that finally matters. In the nineteenth century, the pace of science was not that which Watson describes for twentieth-century molecular genetics, and today we have instantaneous transmission of sounds, words and pictures at a cost even historians can afford.[49]

(4) The final link in the classical diffusion equation is the actual reception by members of the intellectual community in the periphery. Problematic dimensions of the target community are closely related to the discussion above, but a few additional points may be made. Of greatest importance is the fact that the failure of a peripheral community to discuss and focus on particular European ideas may not stem from isolation. Bernard Smith, discussing Australian art, is especially strong on this question.[50] Rejection of a new idea usually has other explanations than isolation. It is not just a question of 'getting it right'. Acceptance, rejection and modification are all part and parcel of what it means to 'receive' a scientific or technological idea.[51]

Finally, the lines usually drawn between popular reception and scientific reception are often very fuzzy indeed. Because of the insistent and ubiquitous interaction between the two domains, separation is sometimes surprisingly difficult and even counterproductive in a historiographical sense.[52] The fine-structured analysis of the reception of scientific ideas outside the scientific centre must firmly base itself in the social and intellectual environment of the scientifically peripheral nation. By this I do not mean to suggest that understanding of central ideas themselves be neglected, but rather that understanding the reconstruction of scientific ideas in the periphery is as historically important as their original construction in the centre.

To illustrate the complexity of the task that awaits the colonial historian, I should like to quote a lengthy paragraph from Glick's new book on relativity reception. After establishing that Einstein's visit had stimulated heady, if uncomprehending, debates about relativity in villages across Spain, Glick imagines for us such a discussion as it might have taken place in a grubby village pharmacy. Beneath the Gogolesque farce, I find in his characters a remarkable dignity. His analysis adds a new dimension to the notion of scientific diffusion both within the scientific community and without.

> The vicar, a political reactionary, has enjoyed an anti-relativity piece in a theological journal, written by a theologian attempting to defend "traditional" values; he opines that Einstein is incomprehensible and has subverted good philosophy and good science with mathematics. The physician, who uses Roberto Novoa Santos' pathology textbook in his daily practice, follows this authority in believing that the time of a patient's reaction to the thump of a neurological hammer bears a "relative" relation to "real" time. The mayor, an alliophile liberal, follows the newspaper in presuming relativity to be a triumph of progress; his authority is a politician, Luis Araquistain, not a scientist. The pharmacist has been assured by Eugenio Pinarua's series in *La Farmacia Espanola* that matter is deprived of its classical meaning by Einstein's theory. When a customer enters the shop and declares that "Light has weight" (the favorite relativity catchphrase of Spanish cartoonists), the village experts will respond in consonance with the most congruent authority. Here we observe relativity diffused through an intellectual field; the variety of social and economic realities ... are likewise refracted through the intellectual field.[53]

In studies of both centre and periphery and of both scientific and lay communities, examples can be given in which the historian has begun to achieve this degree of nuance. Neither the old internalist nor the old externalist mode, followed slavishly, could do it.

As a final point, we must keep in mind the possibility that isolation may at least occasionally greatly benefit the scientific endeavour.[54] Presumably, the argument would be that certain problems in science require the longer, more contemplative, one might even say more 'distanced', view. Or that certain personalities function better in isolation.[55] One thinks of Willard Gibbs walking the quiet streets of nineteenth-century New Haven, William Sutherland in his attic in Dromana, or for that matter Darwin at Down House.

CONCLUSION: TYRANNIES OF THE CENTRE

The myth of 'tyrannical distance' will hold its metaphorical power until Australians walk into their studies and give their globes a spin. As an explanatory tool in the history of science, the concept of distance tells us little about how science works at the periphery. To believe that distance itself kept outlying scientific institutions isolated and undernourished and colonial science subordinate and second-rate is to suggest that the conquering of distance will lead to the establishment of a first-rate and independent scientific establishment in any nation with the will and the wherewithal. But this has rarely happened. In most former colonies, science has remained colonial long after independent political institutions have been erected and long after legions of competent scientists have been put in place.[56] The centre remains 'central', the periphery 'peripheral'. Distance does not tell us why this is so.

The answer lies, I think historians increasingly will agree, in institutional structures and in the power relations that operate between centre and periphery. Political and economic power are important, but power also resides in the scientific infrastructure that serves the centre. In this sense, 'distance' tells us more about the centre than it does about the periphery. From the beginnings of the institutionalization of science and technology in Europe, the centre was learning to initiate, exercise and maintain a network of control over vast distances.[57] The centre learned not only how to cross distances but, even more remarkably, how to control events, people and institutions around the globe.[58] The pay-off for the centre was great: the appropriation of natural resources and natural knowledge from nations that found themselves increasingly dependent economically, scientifically and technologically, even as they fought to establish political independence.

Both science and technology are central elements in the process that produced the first global system of communication and control. In this system, the periphery is as important to the story as is the centre. The difference between the two is the learned ability of the centre to exercise power, authority and control over distance. The scientific centre is not just people or ideas and certainly not just a nation, a region, a continent, or any geographical place. It is rather a network of institutions, an infrastructure that links knowledge to power through the control of the processes of knowledge construction and knowledge communication.

The history of modern science is the account of a new approach to knowledge construction, an approach embedded in social practices utterly different in kind from that of the ancient Greeks, the Arabs, the Yoruba, the Inuit, the Maya, the Chinese, or the Koori. The key to this 'modern' approach is not a magic bullet called 'the scientific method', nor is it the nature or the validity of the knowledge produced. Rather, modern science is distinguished by a set of devices and documents, procedures and techniques, institutions and trained people. From the beginning of modern science, all these elements are employed over distance. And, over time, all these elements are put into place in the furthest periphery. What is maintained in the centre, and what continues to set the centre apart, is the exercise of authority and control.

The history of colonial science, then, must account for the putting into place of this infrastructure (journals, museums, libraries, laboratories, societies, invisible colleges, funding bodies, educational institutions and, of course, people and ideas). The only isolation that limits scientific achievement is the lack of access to these institutions and the lack of position in the hierarchy of authority that stems from these institutions. In the early period of European settlement, Australian scientists suffered from that sort of isolation. However, such isolation was never primarily a function of geographical distance. Australians are not tyrannized by geography but made. The 'distances' that may tyrannize are social, cultural, psychological and finally professional.

All humans and all institutions are in some sense and in some measure constrained by their geographical place. But we must remember that a globe knows no centre and no periphery. Even in science, a 'centre' is defined and constructed by human actors in the service of particular interests. Historically, a centre may be seen as the locus of a particular set of strategies that have proved successful in establishing scientific authority and exercising professional control. Individuals and nations geographically peripheral to such a centre may follow two not necessarily exclusive courses of action: to learn to compete effectively with the aim of rising in the scientific power structure, and/or to develop alternative strategies by redefining the centre.

School of Humanities
Deakin University
Australia

NOTES

* Much of what I say in this paper stems from discussions with my colleagues Barry Butcher, David Turnbull, George Bindon and, at a greater distance, Mark Adams and Sally Kohlstedt.

[1] First published in *Australian Painting Today, The John Murtagh Macrossan Lectures. 1961* (Brisbane, 1962). The lecture was given in the year following the publication of his *European Vision and the South Pacific*, which remains Australia's most seminal contribution to the history of science and of art.

[2] The lecture was recently reprinted in a new anthology of Smith's essays: B. Smith, *The Death of the Artist as Hero* (Melbourne, 1988).

[3] G. Blainey, *The Tyranny of Distance* (Melbourne, 1966).

[4] R. Home, 'The Problem of Intellectual Isolation in Scientific Life', *Historical Records of Australian Science*, 6(1) 1984, 19–30. This thoughtful paper says much of what needs to be said on the issue of isolation in Australian science but seems too ready to attribute the cause of intellectual isolation to geographical distance.
[5] J.G. Jenkin, 'Frederick Soddy's 1904 Visit to Australia and the Subsequent Soddy-Bragg Correspondence: Isolation from Without and Within', *Historical Records of Australian Science*, 6(2) (1985), 153–69
[6] This point has been made by several commentators, most recently by David Knight in his review of the published proceedings of the 1981 Melbourne Scientific Colonialism conference: 'many Britons could feel just as excluded as any Australians or Americans; though with the penny post and the railway system the distances that came to matter were social rather than geographical' (*Historical Records of Australian Science*, 7(1) (1987), 134–6). Knight further develops this notion in his chapter in this volume.
[7] Jenkin, op. cit. (n. 5), p. 165. This paper ably recounts a very interesting episode in Australian science. My quarrel is only with the conclusions the author draws about isolation.
[8] Smith, op. cit. (n. 2), pp. 223, 219.
[9] B. Butcher, 'Darwin's Australian Correspondents: Deference and Collaboration in Colonial Science' in R. MacLeod and P. Rehbock, eds., *Nature in Its Greatest Extent* (Honolulu, 1988), pp. 139–58. In a section devoted to 'Eurocentricity and Australian Science', the author offers a succinct account of the extent to which Australian science was British science through much of the nineteenth century. See also Robert Stafford's chapter in the present volume for a discussion of this theme in relation to the history of Australian geology.
[10] G. Seddon, 'Eurocentrism and Australian Science: Some Examples', *Search*, 12 (1981–82), 446–50.
[11] Canadian science was similarly Eurocentric, though not only English and Scottish but also, of course, French; see Richard Jarrell, 'Differential National Development and Science in the Nineteenth Century', in N. Reingold and M. Rothenberg, eds., *Scientific Colonialism: A Cross-Cultural Comparison* (Washington, 1987), pp. 323–50. Donald Fleming in his famous paper 'Science in Australia, Canada and the United States: Some Comparative Remarks' (*Proceedings of the 10th International Congress on the History of Science, Ithaca, 1962*, pp. 179–96) describes how even the US 'long preserved the residual traces of a colonial posture toward European science' paying 'greater homage in science to France and Germany than to Britain'. Fleming further suggests that Canada was increasingly subject to a 'kind of subinfeudation' to the USA. The fifteen cases studied in the Scientific Colonialism volume (Reingold and Rothenberg, op. cit.) may demonstrate real differences of 'style' in colonial science, yet they are all, almost by definition, in some sense Eurocentric.
[12] See for example, G.G. Joseph, 'Foundations of Eurocentrism in Mathematics', *Race and Class*, 28 (1987), 13–28, for an interesting analysis of the need for historians of science to re-evaluate the indigenous scientific base around the world. Apart from the great interest these traditions exhibit in their own right, they may have played a larger role in many countries than we presently realise in shaping modern understandings of nature. Joseph's paper suggests, for example, a substantial and

complex multicultural influence in the rise of modern mathematics. Furthermore, I would ask why even those indigenous traditions that remained isolated might not be accorded the sort of attention we historians of science have in the past lavished on Leonardo, whose science was also unpublished and largely unrelated to the scientific paradigms and institutions of his time. Also see H. Watson and D.W. Chambers, *Singing the Land, Signing the Land* (Deakin University Press, 1989), in which Aboriginal and European Australian traditions of natural knowledge are systematically and symmetrically compared.

13 Blainey, op. cit., (n. 3), p. ix.

14 The vast literature on Turner is usefully summarised and excerpted in the anthologies *The Frontier Thesis*, ed. R.A. Billington (New York, 1966) and *The Turner Thesis*, ed. G.R. Taylor (New York, 1956).

15 Blainey, op. cit., (n. 3), p. ix.

16 Ibid., p. viii.

17 Would it be naive to suggest that had Blainey felt differently about the matter, he might, with only the slightest of changes to his text, have borrowed Smith's title, *The Myth of Isolation*, and further to suggest that his book, thus entitled, might have engendered a very different response in its readers? One hopes that Australia will never exhibit the jingoism of 'manifest destiny', but must 'destiny and distance' be forever fettered within the Australian psyche?

18 Home, op. cit. (n. 4); Ann Moyal, *'A Bright and Savage Land': Scientists in Colonial Australia* (Sydney, 1986).

19 In 1987 we celebrated the fiftieth anniversary of Robert Merton's 'Science, Technology and Society in Seventeenth-Century England', which showed, among other things, how social origin and educational background limited access to science.

20 Reingold and Rothenberg comment on how difficult it is to decide 'who is at the center and who at the periphery', op. cit. (n. 11), p. xii.

21 S.F. Cannon, *Science in Culture: the Early Victorian Period* (New York, 1978), p. 101. Numbers and Warner quote a similar remark from the *Edinburgh Review* seventeen years later: 'What does the world yet owe to American physicians and surgeons?' (Reingold and Rothenberg, op. cit. [n. 11], p. 191).

22 If Benjamin Franklin allowed some hyperbole to magnify his praise of Rittenhouse as one of the world's best astronomers, nevertheless, he was nearer the mark than those who followed the long tradition of European disparagement of America. One need not multiply examples of this well known phenomenon, which was given a respectable scientific base in the writings of Buffon, but I was recently struck afresh by the mid-18th century comment of the Dutch historian Cornelius De Pauw on the *intellectual sterility* of Cambridge, New England.

23 For example, Reingold, Layton, Kohlstedt, and Numbers, to mention only some of those associated with the 1981 Scientific Colonialism conference.

24 Chambers, 'Two Stages of the Development of Science in Mexico' (PhD dissertation, Harvard University, 1969).

25 Interestingly, Black's comments were added only in the third edition after many of Humboldt's European critics had accused him of seeing Mexico through 'rose-coloured glasses'. A. Humboldt, *Political Essay on New Spain* (London, 1822), p. v–viii.

[26] G.F.FitzGerald to W.F. Barrett, quoted in Jarrell, op. cit. (n. 11), p. 348.
[27] Jarrell, op. cit. (n. 11), pp. 331–6. Jarrell's discussion of the 'asymmetrical' relations of centre and periphery raises some doubts about the traditional assumption of a strictly dominant-subordinate relationship.
[28] Needham was calling Charles Gillespie to account for an absurdly Eurocentric remark. J. Needham, *The Grand Titration* (University of Toronto Press, 1969) p. 54.
[29] Charles Richter, I understand, was an *American* seismologist.
[30] For example, the work of Antonio Alzate, who had no professional scientific appointment, was probably under-rated; on the other hand, Del Rio was apparently respected, but his discovery of a new element (vanadium) was not given adequate attention.
[31] James D. Watson, *The Double Helix* (Harmondsworth, 1970).
[32] On study leave at the University of Melbourne nearly fifteen years ago, I was quite unambiguously told that 'Australia has no intellectual or scientific history'. Thanks to many people attending this conference, we can now say with assurance that, if there was ever a lack, it was of intellectual and scientific historians and not of an intellectual and scientific past.
[33] In Latin American countries the extensive book trade often defied not only distance but the opposition of political and religious institutions, such as the notorious *Index*. See Chambers, 'Period and Process in Colonial and National Science', in Reingold and Rothenberg, op. cit. (n. 11), pp. 306–7.
[34] I have written elsewhere of how Lavoisier's *Elementary Treatise* was translated into Spanish, published in Mexico City within eight years of its appearance in Paris, and immediately put into use as a textbook at the School of Mines. Chambers, op. cit. (n. 33), p. 302.
[35] Shapin, '"Nibbling at the Teats of Science": Edinburgh and the Diffusion of Science in the 1830's', in I. Inkster and J. Morrell, eds., *Metropolis and Province: Science in British Culture, 1780–1850* (London, 1983), pp. 151–78. This is a fine-grained account of scientific diffusion, reflecting the sort of analysis that I find most useful.
[36] The two most interesting collections of this type for our purposes are the cross-cultural studies edited by Thomas Glick, *The Comparative Reception of Darwinism* (Austin, Texas, 1972) and *The Comparative Reception of Relativity* (Dordrecht, 1987).
[37] For a discussion of this approach to innovation and diffusion in the modern context, see, for example, R.F. Rich, ed., *The Knowledge Cycle* (London/Beverly Hills, 1981).
[38] The Arthur Lovejoy approach to the historiography of ideas helped teach us that 'ideas are commodities which cut across interstate lines' and that to account properly for the diffusion of ideas we must ignore traditional disciplinary boundaries not only within the sciences but also across 'an extremely wide range of the philosophical, theological and general literature' over the centuries. Ironically, the history of ideas school went on to become a highly specialized field with its own fairly rigid boundaries. See the first essay in A.O. Lovejoy, *Essays in the History of Ideas* (Baltimore, 1948) and of course his *The Great Chain of Being* (Cambridge, Mass., 1953).

39 One example that illustrates all of these points is the wonderful story of the European 'discovery' of the platypus. See H. Burrell, *The Platypus* (Adelaide, 1927), and K. Dugan, 'The Zoological Exploration of the Australian Region and Its Impact on Biological Theory', in Reingold and Rothenberg, op. cit. (n. 11), pp. 79–100. Another rich episode is the fight by Ferdinand Mueller to author the *Flora Australiensis* and his ultimate defeat by the experts at the centre. Analytically, this struggle holds much of relevance to the understanding of the role of authority in centre-periphery relations. See the interesting account in Ann Moyal's highly readable survey of scientists in colonial Australia, op. cit. (n. 18), pp. 149–53.

40 The virtual absence of these channels to most Islamic and African countries is explanation enough for their failure to 'receive' Darwinism, quite apart from the many cultural reasons that might be offered.

41 Roberto Moreno, 'Mexico', in Glick, op. cit. (n. 36; 1972), p. 348.

42 Gary S. Dunbar, '"The Compass Follows the Flag": The French Scientific Mission to Mexico, 1864–1867', *Annals of the Association of American Geographers*, 78 (1988), 229–40.

43 Enrique Beltran, 'La Science Française au Mexique', *Culture Française*, 9(4) (1960), pp. 9–22.

44 Butcher, op. cit. (n. 9). This subtle and very useful analysis, which never takes recourse to 'distance' in its explanations, throws much light on Australian biology during the period. For example, Butcher makes a very important distinction between 'deferential' scientific workers (gatherers in the classic Baconian mode) and 'collaborative' scientific workers (hunters in the Humboldtian mode).

45 Frederick Burkhardt in Glick, op. cit. (n. 36; 1972), p. 33.

46 Home, op. cit. (n. 4), p. 20.

47 And I write these words staring out over the Southern Ocean, reasonably far removed even from Melbourne.

48 G. Davison, *The Rise and Fall of Marvellous Melbourne* (Melbourne, 1979), p. 131. Stephen Alomes has written on questions of colonial dependency with regard to Australia in 'The Satellite Society', *Journal of Australian Studies*, 9 (1981), 2–20.

49 Ironically, today, journals travelling by sea mail once again take at least fourteen weeks.

50 Smith, op. cit. (n. 2), p. 225.

51 We are fully prepared nowadays to accept the idea that a technological innovation from the industrial centre may or may not fit, with or without modification, into an industrially peripheral environment. We must have the same courage to accept the notion that ideas, in some cases more than others, may be as culture-bound and as culturally inappropriate as certain technical devices. For a first-rate case study of this process in technology, see Edwin Layton, 'European Origins of the American Engineering Style of the Nineteenth Century', in Reingold and Rothenberg, op. cit. (n. 11), pp. 151–66.

52 In many other cases, of course, the public may little note nor long remember what scientists have to say.

53 Thomas A. Glick, 'Relativity in Spain', Glick op. cit. (n. 36, 1987), p. 255.

54 Such a possibility has been mentioned by Home and other writers. Barry Jones, current Australian Minister for Science, recently suggested that the 'tyranny of

distance … was a spur to innovation and inspired improvisation': see *In Future*, 9 (7) (1988), 7.

55 I believe that Emerson referred to 'that necessity of isolation that genius feels'.

56 Chambers, op. cit. (n. 33), p. 314.

57 Bruno Latour, *Science in Action* (Milton Keynes, 1987). See especially Chapter 6, 'Centres of Calculation', which is another work that helps set the new agenda for historians of colonial science.

58 John Law, ed., *Power, Action and Belief* (London, 1986). This anthology also has much of interest to the historian colonial science. For present purposes, see Law's paper, 'On the Methods of Long-distance Control: Vessels, Navigation, and the Portuguese Route to India', pp. 234–63.

DAVID KNIGHT

TYRANNIES OF DISTANCE IN BRITISH SCIENCE

One of the major themes in this collection is distance in science, and the problems and opportunities presented by dependence, independence and interdependence; another is the relationship of the central or metropolitan and the peripheral. These things are easy to see in the context of the USA, politically but not yet culturally independent in the nineteenth century, and Australia, still formally a colony into the twentieth century; but they can also be seen within Britain, in the relationships between those in or near London (especially in Oxford and Cambridge) and those in the provinces. The position of somebody in the north of England, like John Dalton, was not so very different from that of a colonial; when he bade farewell to his audience after a course of lectures in London to return to 'comparative retirement' in Manchester he sounded like someone going home to a colony.[1] The Australian or American experience of being snubbed or patronized happened to Englishmen all the time; and the sort of people who wrote disparagingly about the domestic manners of Americans or Australians reacted in the same way to those of manufacturing districts in their own country. This indeed is classic metropolitan behaviour.

Australia did not loom very large in British consciousness throughout the century, unlike Ireland which rocked Church establishment, underwent famine, produced terrorism and split a great political party. Britons were happy when it suited them to applaud the progress made by their Anglo-Saxon cousins in subduing the wilderness and replacing savage races, regarding both Americans and Australians as honorary Englishmen of some kind. There was a steady flow of emigrants to both countries, more going to the USA; but the cultural centre for these people, however much they may have disliked the country they had left, remained Britain. For intellectuals like Henry James and Ernest Rutherford, south-east England was the place to go. The British system, if one can dignify it by such a term, of supporting science through patronage, and breaking into tight groups, meant that few American scientists (after the efflux of

R.W. Home and S.G. Kohlstedt (eds.), International Science and National Scientific Identity, 39–53.

Loyalists at Independence brought W.C. Wells and Rumford to London) came to Britain permanently. Australians and Mancunians found it easier, but heterodoxy made it difficult as we shall see with William Swainson, and Dalton like some colonials preferred to be a big fish in a small pond. We shall look at Humphry Davy, leaving Cornwall to make a career in London, receiving and then dispensing patronage; and at Swainson, whose bid for metropolitan recognition did not in the end come off and who emigrated in a huff to the colonies. Neither of them was particularly happy in the end, but that is perhaps less our concern.

Britain is a small country, but this was less obvious in the days before steamships and railways. A geologist in Devonshire[2] or a chemist in Manchester could seem very distant from those in London, Cambridge or Edinburgh; and geographical isolation could be a real problem. This was one of the ills that the British Association for the Advancement of Science, founded in 1831 just on the verges of the railway age, was meant to relieve.[3] By meeting each year in a different provincial city, the Association did give everybody from time to time a chance to see and hear the eminent, perhaps to be seen and heard, and at least to meet the likeminded. Even now, the north-east of England seems a long way from the centre of things; for the nineteenth century it is not easy to distinguish robust provincial pride from protesting too much about impotence and distance. One reason is that social and geographical distances go together.

Londoners then and since take it for granted that everything important happens there,[4] though in intellectual life Oxford and Cambridge might count as extensions of London. Provincials are not taken seriously unless they settle down and become Londoners, while for anybody brought up there it even remains difficult to work on scientists of the past who never left the provinces. Provincials and their institutions are perceived as unsophisticated and slightly comic; they might be expected to feel inferior, and perhaps bluster about it. This is a tradition that goes back to Shakespeare; and the achievements of Manchester in the last century cut little ice in London. To the ambitious in the nineteenth century, London was a magnet. There it first became possible to make a career for oneself in science, as Davy, Faraday, Tyndall, Huxley and Crookes did. Here were the Royal Society and the Royal Institution, the great museums, Greenwich Observatory, and the centre of government not merely of an island but of an empire. London was the Rome or Constantinople of the nineteenth century, anyway to Londoners; to be tired of it was to be tired of life. Cobbett saw it as the great wen, an excrescence or ulcer draining the life blood of Britain. Others hated its fog and filth. But it was the place to be for anybody who wanted to keep up and get on, especially in the intellectual world: and in this world the nineteenth century was the age of science.

An intellectual distance, and a difference in outlook, separates Londoners from those living further away. Equally important, however, was difference in social rank. Gentlemen who had estates in the country came up to London for the season, spending the summer and early autumn at home, and the meetings of the Royal and Linnean Societies (and the other specialised societies that came to join them as the century wore on) were arranged to fit in with this programme.[5] Science in Britain in the early nineteenth century was an occupation for gentlemen, after all, or for those in the learned professions – clergy, physicians or lawyers – who with a degree from an ancient university had the right to describe themselves as Esquire and thus to acquire the status of gentlemen. Even at Oxford and Cambridge, there were great social distances between the gentlemen commoners, there for a finishing school, and those from the less affluent classes, mostly training to be clergymen.

Nationalism does not seem to have been especially important for science in nineteenth-century Britain, especially because France at first and then Germany was clearly the centre of things and Britain always rather an outsider; it is after all an offshore island. Only from time to time, as with geology in the 1820s and '30s and electromagnetism later, were Britons in the lead. British inferiority was often manifest, especially perhaps in chemistry and physiology, and a rhetoric of decline played an important part not only in the first years of the BAAS but throughout the century. Patriotism is a different thing. It was not science that gave the English an image of themselves, but the splendid Constitution issuing out of the Glorious Revolution of 1688, imitated fairly well by Brother Jonathan in 1776 (though American democracy was often seen as a dreadful warning) and very badly by the French in their overrated revolution of 1789. Although aristocrats in Britain feared a revolution on the French model, especially in the thirty years after Waterloo, gradual political changes, organised religion and economic prosperity combined to ensure political stability. Regimes without the long standing of British governments, and without natural frontiers, might promote a nationalism based upon scientific achievements to bring them legitimacy. There was no great need in Britain to promote those episodes where Britons had done exciting things in science – though nobody was allowed to forget that Newton was an Englishman. Germans, coming like Scots from a poorer country with a better educational system than that of England, played important roles in pure and applied science[6] in the late nineteenth century in their adopted country, just as they were very important in Australasia. The English exported to Australia a pattern of scientific activity in which social distinctions were extremely important and social distances preserved, but not one in which foreign links were to be despised or rejected.

Those living in provincial towns in England found, and still find, what must also be familiar to Australians; that the most eminent intellectuals who grew up there went to the London area to make their career. Only around London was there a big enough intellectual community. This was true also for those from the other nations within the British Isles: in London and Cambridge, Irishmen like Tyndall and Stokes and Scots like Brown, Maxwell and Ramsay all settled down, just like those from the English provinces.[7] Snobbery is more important than nationalism in structuring British science in the nineteenth century and except for the few of independent means, most scientists depended upon patronage, especially in the first half of the century. Recognition abroad might help, early on (as for Davy) from France, and later, especially in chemistry, from Germany; and those of unsure status might therefore be more cosmopolitan. No Englishman in the interests of nationalism ever seems to have thought of refusing membership of an overseas academy, as Rod Home suggests Australians might more recently have done; indeed, they generally listed such things on their title-pages. Whether coming from a colony was worse than coming from the industrial North is an interesting question, to which it would be very hard to give a definite answer in general terms.

Banks had made his reputation with his voyage to Botany Bay, but his status depended upon his land and family; the two together led to his Presidency of the Royal Society from 1778 to 1820.[8] During this time he exercised patronage, notably in voyages of exploration such as that of Flinders to Australia, and in the wool trade, but also at home in the Board of Agriculture, where he promoted Davy's career. Davy had grown up in Cornwall, then a backward corner of the country, part of the Celtic fringe, where his father had been a wood-carver when in work and his mother when widowed a milliner. In the manner characteristic of the day, the Davys liked to think of themselves as a good family fallen on bad times; but Henry Mayhew was one of those struck by the social mobility Davy achieved through science.[9]

With the patronage first of James Watt, whose son Gregory had stayed with the Davys in the hopes of curing his tuberculosis in the mild climate of Penzance, and then of other members of the Lunar Society when he worked as assistant to Beddoes[10] at Clifton, Davy was propelled with the success of his work on the oxides of nitrogen to a post at the Royal Institution in London. His lectures here, including a series on agricultural chemistry, and then his electrochemical research, made him in the course of a few years one of the best-known scientists of the day; and his Bakerian Lecture to the Royal Society of 1806 was crowned by the award of a prize by the Institut in Paris. He became a great diner-out, and was clearly a spirited talker; and he put behind him his apothecary's apprenticeship in rural Cornwall, which he hardly ever revisited. He was happy to be in the society of aristocrats but relaxed only in the company of professional men, particularly doctors, with whom he went fishing.[11] Social mobility has a high price in unease.

His first Continental trip, to Napoleon's Paris to receive his prize, was followed by others. He was abroad when he heard of Banks' fatal illness, and hurried back as a candidate for the Presidency of the Royal Society: Banks had been elected a few days before Davy's birth. Banks and others had hoped that Wollaston would succeed him, but Wollaston withdrew rather than face a contest with Davy, who emerged as the only serious candidate. It was astonishing that one from his background should have risen to a position of such eminence, and some of his elders and betters were appalled. That the part of Banks should now be played by Davy, who had a reputation as a philistine bored by the pictures in the Louvre and whose cocked hat was usually askew, was mildly shocking.

In the event, Davy's Presidency was not a great success. His own research on protecting the copper bottoms of ships from corrosion was unsuccessful, and the expectations of the hostile camps[12] between which he found himself could not be satisfied. He tried to keep up the style of Banks, having acquired the means by an unhappy marriage to a wealthy widow, but found social climbing a lonely activity. A president neither born to the purple nor a university graduate was not in an easy position socially, and Davy's fast-working mind, his rhetorical skills and his conviction of his scientific merits did not necessarily endear him to everybody. Even with Faraday, to whom he became a father in science, he proved an oppressive parent: Faraday learned from Davy's example to avoid exposed positions of great eminence, though he followed Davy's research programme with even greater success than Davy himself.[13]

One might have expected that a President with very different interests from Banks's might have propelled British science in a different direction, towards the laboratory rather than out into the Antipodes. But during Davy's time, voyages of exploration in the tradition of Cook and Banks continued, especially to the Canadian Arctic and to South America. And just as Banks had spent much time and effort in acclimatising the Merino sheep in Britain, so Davy was with Raffles a founder of the London Zoo. He hoped that some exotic creatures might be made economically important in Britain; but English grasslands were never graced with alpacas or kangaroos and the Zoo became a place where, as at the Royal Institution, entertainment and scientific study went on in uneasy union.

Bank's position had made it easy for him to exercise patronage, though he missed out on some promising men – for instance Faraday. Patronage is bound to be hit and miss, perhaps like any kind of selection. Davy used his patronage to obtain for his friend J.G. Children a post in zoology at the British Museum.[14] Children did not know any zoology but was a mineralogist and chemist of independent means, who had worked with Davy on gunpowder and also on the vile-smelling hydrogen telluride. When he fell upon hard times, Davy secured him a job, which Children did not at first particularly want, at the Museum. In the event he proved a competent zoologist, but the British Museum was not transformed as Davy had hoped it might be into something resembling the Musée d'Histoire Naturelle in Paris; that only happened when the natural history collections were moved to South Kensington in the 1880s,[15] under Owen who had strong Australasian connections. Children's social status was such that he could deal with the Museum's Trustees as an equal rather than as an employee, which was an advantage: but his appointment was controversial, and the prestige of the Museum was dented by the attacks of an unsuccessful candidate for the post, William Swainson, and his friend Thomas Traill. Collections made on national expeditions were sometimes sent elsewhere, like Darwin's from HMS *Beagle*. Mismanaged patronage seriously damaged an institution that Banks had successfully promoted.

Davy's successor in 1827 was the Cornish MP, Davies Gilbert (formerly Giddy); but after a brief reign he stood down and there was a bitterly contested election between John Herschel and the Duke of Sussex, the most intellectual of Queen Victoria's uncles. This raised in acute form questions of rank, professionalism and the status of science in general and the Royal Society in particular.[16] Even an affable Royal Duke was unlikely to become the close friend of working scientists but his brother, soon to be King William IV, had as Lord High Admiral sponsored scientific voyages, and the Duke himself took a genial interest in the advance of knowledge. He won the election; and we get a clear picture of social distance in letters written to Swainson by Peter Mark Roget, a Secretary of the Society (the other being Children) and author of the famous *Thesaurus*.[17] Roget was a doctor, and was responsible for the new journal that became *Proceedings of the Royal Society*, less formal than its *Philosophical Transactions*; he also wrote a Bridgewater Treatise on the goodness and wisdom of God.

The Duke had been elected on 30 November 1830, and on 5 December Roget described his first 'interview (I should perhaps say an audience) with our new President'. Clearly, the Duke succeeded in charming Roget, who earnestly wished 'he may eventually succeed in obtaining the cooperation of those who now assume so hostile an attitude, & threaten to secede from the Society, & perhaps establish another of their own'. His misgivings were less about the Duke than about the 'very imbecile Council' elected at the same time, because those who 'arranged' it 'evidently took pains to retain those of the former Council who were the most inefficient members, & particularly those who have never once attended: while on the other hand they excluded the really valuable members, & put into their places persons who, with one or two exceptions, will be totally inactive'.

By 2 March 1831 Roget was finding the Duke less assiduous than he might have been: 'he has not yet commenced either his evening or his morning conversazione's (sic)' and 'has been very little with us of late: & tomorrow is the first day, since the 20th of January, that he will take the chair at the meetings: having been during almost the whole of the interval at Brighton', where his brother George IV had built a palace of more than oriental splendour. To Roget's disappointment, the friends of Herschel did not respond to ducal overtures and the prospect of a return to the days of Banks when science had basked in the approval of the Establishment: 'The Duke takes every means in his power that is at the same time consistent with the dignity of his office, to conciliate them: but as yet without the smallest success', Roget wrote on 12 March 1831. The President had proposed revision of the Society's Charter, and 'We have taken care to nominate on that committee the principal reformers, both moderate and radical: in order that their plans may undergo full & fair discussion. Of this number are Babbage, Baily, Beaufort, Brown, Fitton, Herschel and Warburton. These have, by letter, declined attending, without assigning any reason: and their places must be filled up by others. This is, indeed, to use your words, "carrying matters much too far"'.

Roget still hoped 'that the anticipated good from having a Royal President may eventually arrive'. In December 1830 he had indeed felt that if party spirit were to go on running so high, he would resign as Secretary in November 1831; but in the event he continued until 1849, by which time the Royal Society was transforming itself from a club into an Academy of Sciences. What the episode shows is that considerations of rank and status were closely involved with the moves towards specialisation and professionalisation, and that social distances were formidable, especially just at the time of the great Reform Bill of 1832.

Roget's letters were to William Swainson, a man who failed to cross these distances and who as a consequence later sailed as an emigrant to New Zealand, and for a time worked in Australia. Swainson did not have so far to travel as Davy or Faraday, who came from the artisan class. He came from Liverpool where his father had been a Customs Officer, and during the Napoleonic Wars he served in the Mediterranean with the Commissariat. On his return, after a voyage to Brazil, he took advantage of the new process of lithography as a cheap way of printing natural-history plates, especially of birds and shells. He was a very talented artist and supported himself and an increasing family by publishing illustrations, and by buying and selling specimens. We can learn the details from his correspondence.[18] But Swainson rated an illustrator some distance below a naturalist, and longed to shine as a writer.

His chance came with government support for publication of the natural history of the northern voyages, and then with the publishing revolution of about 1830, associated with the 'March of Mind', that brought cheap case-bound books to the British public. Swainson undertook to write the natural history for Dionysius Lardner's series, *The Cabinet Cyclopedia*, a series of informative volumes of which the first, Herschel's *Preliminary Discourse*, is the best-known.[19] Swainson wrote a *Preliminary Discourse* himself, on natural history, which is interesting for its descriptions of the generally rather turgid meeting practices of scientific societies, and of natural history publishing. In this and his other works he adopted the Quinary System of classification, an attempt to find an order that was both natural and tidy. He much admired the French and saw 'decline' in British science, but he did not follow Cuvier's lead where theory was concerned. His attempt to close the distance between the artist and the theorist did not work, because his persistence in unorthodoxy (the quinary system went with his strongly high-church views) kept him apart from most zoologists; McLeay, the other notable quinarian, being in Australia. But Swainson was also not quite a gentleman born and he was of a very quarrelsome disposition, suitable perhaps to a generation that had grown up in a war so long that peace after 1815 seemed abnormal and military metaphors natural.

The Quinary System never lost its heretical taint; but small groups that have had to fight for status and for truth as they see it remain defensive and distanced even after victory. Thus the Darwinians who formed the X Club, some like Huxley and Hooker with Australian connections, seem to have seen themselves as an embattled minority within the 'Church Scientific' in their struggles to cut British science free from religious ties and amateurism.[20] It is difficult here to separate social and intellectual distances; although the men of the X Club held the highest offices in science, like Davy they never felt easy in the Establishment. He tried to be a Gentleman of Science, as Faraday and the X Club members did not; but nobody in nineteenth-century Britain (and to some extent still) could avoid questions of social rank and the distances associated with it.

One way in which this showed itself was in attitudes to science as a profession. In writing about Faraday, Tyndall referred to his giving up professional work (meaning routine analyses for a fee) in favour of research into electricity.[21] Davy had refused in a gentlemanly way to patent his safety lamp, comparing himself favourably to Wollaston, born a gentleman, who had made a great deal of money from work with platinum. In Davy's time, pure and applied science were not distinguished: science was opposed to practice, or mere rule of thumb.[22] By the second half of the century, however, a gulf had opened, especially in chemistry: professional chemists earned their living generally by analyses, while academic or learned chemists increasingly began to find posts in the expanding educational system especially after 1870. The distance between the two groups, both social and intellectual, led to a schism in the Chemical Society of London, which continued as a learned body while the Royal Institute of Chemistry represented the professional interests of those in industry, whose status in the nineteenth century remained low. It was a century before this disruption was healed.[23]

By this time there were also distances within the scientific community: the sciences not only had a pecking-order changing over time, but also became increasingly specialised. Davy, Wollaston and Young worked across a spectrum of sciences. Herschel was one of the last who could refuse to specialise,[24] and by the end of the century most practitioners saw themselves not as men of science but as, for example, organic chemists. To Rutherford, 'chemist' meant damn fool; the remark had a particular context, but it reminds us of the distances between the sciences that had become a feature of the specialised life of the nineteenth century.

In the 1880s the *Proceedings of the Royal Society* were divided into parts A and B, A including the physical sciences and B what my physics teacher used to call 'the less exacting discipline of the more descriptive sciences'. Journals like *Nature* were required to publish each week the latest work and also to keep those in one specialism aware of what was going on in others, something that in more leisurely days Mary Somerville, Herschel and others had done in books or in expansive essays in the *Edinburgh*, the *Quarterly* and other *Reviews*.[25] Crookes had tried, with *The Quarterly Journal of Science*, to make a scientific review work, but the forces of specialism were too strong. To see 'two cultures' in nineteenth-century Britain is too simple. There were divisions not just between scientists and those brought up in the humanities (overlapping groups at the academic end) but also between different kinds of scientists, with a ranking order. The social history of science is a complicated business, perhaps especially in England where amateurism continued longer than in France and Germany, and where the class system seems to have been maintained with unusual and particular care. Democratic Americans love to rank scientists as quantitatively as possible; the British had, and maybe have, more subtle and probably more offensive ways of doing it.

Australia was very distant from Britain geographically, but not necessarily in other ways; it still seems very British. Banks and Darwin both went there, and for neither of them were social distances a problem. Their visits can be seen as exemplifying the Baconianism so much invoked in nineteenth-century Britain.[26] Bacon's most accessible work, the fragmentary *New Atlantis*, had after all described a small island from which fact-collectors went out and brought back information, to be reasoned upon by those further up the social pyramid. Banks's move to the administration of science and Darwin's into theorising fit this pattern, for one need not remain in the fact-gathering stage throughout life. Visiting Australia was thus a stage in a career: the fauna and flora of Australia, the stars of the southern hemisphere, or the behaviour of wombats in the wild, cannot be observed in London.[27] There is some science that must be done in Australia, just as there was some that had to be done in the provinces: the Silurian and Cambrian rocks had to be studied in Wales,[28] and pendulums swung down mines in Cornwall and in Durham as well as in Tasmania in studying gravitation.

The reasoning upon and publication of these and other pieces of 'Humboldtian' science was done, as Humboldt had done it, in great European centres, where there were scientific societies, museums and libraries. This pattern was extended to the British Association: those in the provinces might furnish, especially in natural history and astronomical observation, material for those in the metropolis to think about. Londoners may well have felt that Australia was no more distant intellectually than the industrial North, and certainly that the science done by gentlemanly visitors and the colonial administrators who supported them was of central importance. Right through the nineteenth century reputations were made on scientific voyages, and important posts might follow. Not only Banks but also Sabine, Hooker and Huxley became President of the Royal Society, and the voyage of HMS *Challenger* was one of the most important bits of big science done in late Victorian Britain. This also involved scientists from a number of countries in its writing up, in a self-conscious repudiation of nationalism.

The problems came, no doubt, where geographical distance was reinforced, in sciences that did not need to be done in Australia and that in the British empire were already being done at home, for example in physics and chemistry in their more abstract branches, specialised disciplines where even in Britain the communities were not large. In such fields one might be worse off in Australia than in Japan,[29] for which there was no metropolitan cultural centre at a distance; and one would certainly be worse off than those in the provinces in Britain, who had a rapid penny postage by the second half of the century and speedy train services making visits to London or to BAAS meetings practicable. Australian physical science could not but be provincial, and those thinking of teaching science in Australian institutions had to bear that in mind. Symbolic visits to the colonies by the BAAS every twenty years or so were no substitute for easy contact. Science in Britain was affected by all sorts of distances in the shifting pattern of class-consciousness in the nineteenth century: the empire presented opportunities to Hooker and Sabine, Rutherford and Bragg, but to others it must have closed doors and made the centre of things feel very far away.

Australians probably also felt that they were treated as second-class citizens in terms of resources available, as colonials to be exploited.

Australians and Americans use the word 'colonial' in a rather different way from everybody else. Their experience, except for those who are Aborigines or American Indians, has been so very different from that of Indians or Nigerians that it is misleading to use the word 'colonial' for both situations; 'provincial' seems a better way of describing science in the former case. A recent analysis of the British empire in the second half of the nineteenth century[30] confirms the British impression that running an empire was expensive and that those on the periphery did best out of it, an impression that Americans looking back on the years since 1945 might share. The close links between home and colonies that one imagines do not show up in economic terms, and Australian education like Australian defence seems to have been indirectly supported by the taxpayer in Britain; Australians are probably right in thinking theirs has been a lucky country.

Whether jumping out of the British fire into the American frying pan would make life more comfortable for Australian scientists remains to be seen; the Canadian experience here[31] might well be instructive. What seems certain is that the *New Atlantis* model is inevitable, whether one is in Melbourne or in Durham, until the local scientific community can somehow grow large enough to be self-sustaining. Then patriotism is not enough – in nineteenth-century Britain contact with men of science abroad gave the chance to escape from the stultifying effects of social distances – though in the last resort one had to work within the social system in which one found oneself. The British experience seems to be that a multiplicity of international connections helps science grow. For different people at different times and in different disciplines, important contacts in the nineteenth century were not just with France or Germany, but also with Switzerland, Sweden, Belgium, Russia, Italy and the Netherlands. These geographically-distant connections were sometimes easier than socially-distant ones within the same city; for as well as being a period of great class-consciousness, the Victorian era in Britain was also one of momentous rows. Distances of different kinds – geography, language and status – may tyrannise science, or they can be a great resource of variability. Australian science may have suffered because Britain was more important to Australia than the other way round; but to the outsider it looks as though it had quite a flourishing growth.

Department of Philosophy
University of Durham
U.K.

NOTES

1 H.E. Roscoe and A. Harden, *A New View of the Origin of Dalton's Atomic Theory* (London, 1896), p. 122.
2 M.J.S. Rudwick, *The Great Devonian Controversy: The Shaping of Scientific Knowledge among Gentlemanly Specialists* (Chicago, 1985); C.A. Russell, *Science and Social Change, 1700–1900* (London, 1983); D.M. Knight, *The Age of Science* (Oxford, 1986, new. ed. 1988).
3 J. Morrell and A. Thackray, *Gentlemen of Science* (Oxford, 1981); and the associated *Correspondence* (London, 1984); R. MacLeod and P. Collins (eds.). *The Parliament of Science* (London, 1981).
4 L. Stone, *The Family, Sex and Marriage in England, 1500–1800* (new ed., Harmondsworth, 1982), p. 179. I. Inkster and J. Morrell (eds.), *Metropolis and Province* (London, 1983); I. Inkster (ed.), *The Steam Intellect Societies* (Nottingham, 1985). For earlier London dominance, see P. Earle, *The Making of the English Middle Class: Business, Society and Family Life in London, 1660–1730* (London, 1989). On provincial geology, see H.S. Torrens, 'Arthur Aitken's Mineralogical Survey of Shropshire 1796–1816, and the Contemporary Audience for Geological Publications', *British Journal for the History of Science*, 16 (1983), 111–53.
5 S. Forgan, 'Context, Image and Function', *British Journal for the History of Science*, 19 (1986), 89–113, looks at the buildings and arrangements of scientific societies.
6 On British geology at a triumphal period, see Robert A. Stafford's paper in this volume. On Albert Gunther, see A.E. Gunther, *A Century of Zoology at the British Museum* (London, 1975); on an industrialist, S.E. Koss, *Sir John Brunner: Radical Plutocrat, 1842–1919* (Cambridge, 1970). On Mueller, see R.W. Home's paper in this volume; it is interesting that the telescope at the Sydney Observatory came from Germany. See also A.M. Lucas, 'Baron von Mueller: Protégé turned Patron', pp. 133–52 in R.W. Home (ed.), *Australian Science in the Making* (Melbourne, 1988), a very useful collection of papers giving an excellent overview and some close studies.
7 Two important scientists who remained in Ireland were Lord Rosse and Hamilton; see T.L. Hankins, *Sir William Rowan Hamilton* (Baltimore, 1980).
8 H.B. Carter, *Sir Joseph Banks* (London, 1988). On Brown, D.J. Mabberley, *Jupiter Botanicus* (Braunschweig, 1985).
9 H. Davy, *Collected Works*, ed. J. Davy, Vol. 1 (London, 1839–40); H. Mayhew, *The Wonders of Science, or Young Humphry Davy: The Life of a Wonderful Boy written for Boys*, 2nd ed. (London, 1856).
10 T.H. Levere, 'Dr Thomas Beddoes; Science and Medicine in Politics and Society', *British Journal for the History of Science*, 17 (1984), 187–204; D.A. Stansfield, *Thomas Beddoes* (Dordrecht, 1984).
11 S. Forgan (ed.), *Science and the Sons of Genius: Studies on Humphry Davy* (London, 1980); my paper on fishing is on pp. 201–30.
12 D.P. Miller, 'Between Hostile Camps; Sir Humphry Davy's Presidency of the Royal Society, 1820–1827', *British Journal for the History of Science*, 16 (1983), 1–47. H.B. Carter (ed.), *The Sheep and Wool Correspondence of Sir Joseph Banks, 1781–1820* (Sydney, 1979).

[13] D. Gooding & F. James (eds.), *Faraday Rediscovered* (London, 1985); my paper on Davy and Faraday is on pp. 33–49.
[14] A.E. Gunther, *Founders of Science at the British Museum, 1753–1900* (Halesworth, 1980).
[15] W.H. Flower, *Essays on Museums* (London, 1898); M. Girouard, *Alfred Waterhouse and the Natural History Museum* (London, 1981). On Owen, see the paper by Elizabeth Newland in this volume and, for a more sympathetic account, Jacob W. Gruber, 'From Myth to Reality: the Case of the Moa', *Archives of Natural History*, 14 (1987), 339–52. On the row over Children's appointment, see A. Gunther, 'President's Anniversary Address', *Proceedings of the Linnean Society*, 112 (1899–1900), 14–24, pp. 19f.
[16] M.B. Hall, *All Scientists Now: The Royal Society in the Nineteenth Century* (Cambridge, 1984); and see the informal writings of the Society's Assistant Secretary, W. White, *Journals* (London, 1898).
[17] Gunther's address (n. 15) describes Swainson's correspondence preserved at the Linnean Society, and is followed (pp. 25–61) by a summary Calendar arranged under correspondents. Incoming letters only are preserved.
[18] D.M. Knight, 'Ramsbottom Lecture: William Swainson, Naturalist, Author and Illustrator', *Archives of Natural History*, 13 (1986), 275–90; S. Natusch and G. Swainson, *William Swainson; the Anatomy of a Nineteenth-century Naturalist* (Wellington, NZ, 2nd printing 1987); this reproduces the Calendar of Swainson's correspondents, and lists other MS sources; G.M. Swainson (ed.), *William Swainson: Diaries, 1808–1818* (Palmerston North, NZ, 1989). On the trade in specimens, see M.A. Taylor and H.S. Torrens, 'Saleswoman to a New Science: Mary Anning and the Fossil Fish *Squaloraja* from the Lias of Lyme Regis', *Proceedings of the Dorset Natural History and Archaeological Society*, 108 (1986), 135–48.
[19] J.F.W. Herschel, *Preliminary Discourse to the Study of Natural Philosophy* (London, 1830, reprint introduced by A. Fine, Chicago, 1987). J.N. Hays, 'The Rise and Fall of Dionysius Lardner', *Annals of Science*, 38 (1981), 527–42. D.M. Knight, *Zoological Illustration* (Folkestone, 1977); A. Ellenius (ed.), *The Natural Sciences and the Arts* (Uppsala, 1985).
[20] R.V. Jensen, 'Return to the Wilberforce-Huxley Debate', *British Journal for the History of Science*, 21 (1988), 161–79; Ruth Barton, ' "An Influential Set of Chaps": the X-Club and Royal Society Politics, 1864–1885', *British Journal for the History of Science*, 23 (1990), 53–81.
[21] See my paper on Faraday in R. Porter (ed.), *Man Masters Nature* (London, 1987), pp. 126–36; J. Tyndall, *Faraday as a Discoverer* (London, 1868).
[22] R.F. Bud and G.K. Roberts, *Science versus Practice* (Manchester, 1984). P. Alter, *The Reluctant Patron: Science and the State in Britain, 1850–1920* (Oxford, 1987).
[23] C.A. Russell, N.G. Coley and G.K. Roberts, *Chemists by Profession* (Milton Keynes, 1977). Crookes' journal *Chemical News* reached both professional and learned chemists, and its Letters to the Editor in the 1870s are most interesting on the distances between them.
[24] S.F. Cannon, *Science in Culture* (New York, 1978), esp. chapters 2 and 3.
[25] A.J. Meadows (ed.), *Development of Science Publishing in Europe* (Amsterdam, 1980); W.H. Brock and A.J. Meadows, *The Lamp of Learning* (London, 1984).

[26] R. Yeo, 'An Idol of the Market-place: Baconianism in Nineteenth-century Britain', *History of Science*, 23 (1985), 251–98.
[27] A. Moyal, *'A Bright and Savage Land': Scientists in Colonial Australia* (Sydney, 1987).
[28] J.A. Secord, *Controversy in Victorian Geology: The Cambrian-Silurian Dispute* (Princeton, 1986); and 'The Geological Survey of Great Britain as a Research School, 1839–1855', *History of Science*, 24 (1986), 223–75; N.A. Rupke, *The Great Chain of History: William Buckland and the English School of Geology, 1814–1849* (Oxford, 1983).
[29] R.W. Home, 'Physics in Australia and Japan to 1914', *Annals of Science*, 44 (1987), 215–35; idem, 'The Problem of Intellectual Isolation in Scientific Life: W.H. Bragg and the Australian Scientific Community, 1886–1909', *Historical Records of Australian Science*, 6 (1) (1984), 19–30, and 'First Physicist in Australia: Richard Threlfall at the University of Sydney, 1886–1898', *ibid.*, 6 (3) (1986), 333–57.
[30] Patrick K. O'Brien, 'The Costs and Benefits of British Imperialism, 1846–1914', *Past and Present*, 120 (1988), 163–200, and the ensuing debate between O'Brien and P. Kennedy, *ibid.*, 125 (1989), 186–99.
[31] Trevor H. Levere, 'The History of Science in Canada', *British Journal for the History of Science*, 21 (1988), 419–25. On the organisation of science in Australia in the last hundred years, see R. MacLeod (ed.), *The Commonwealth of Science: ANZAAS and the Scientific Enterprise in Australasia, 1888–1988* (Melbourne, 1988).

ELIZABETH DALTON NEWLAND

DR GEORGE BENNETT AND SIR RICHARD OWEN:

*A Case Study of the Colonization of Early Australian Science**

> It is astonishing with what intense feeling of hatred Owen is regarded by the majority of his contemporaries ... The truth is, he is superior of most, and does not conceal that he knows it, and it must be confessed that he does some very ill natured tricks now and then.[1]

By the early nineteenth century, European naturalists had become increasingly aware of the uniqueness and significance of the flora and fauna of the new British colony in the South Pacific, the colony of New South Wales. The distinctive plants and animals had an immense impact on the biological thought of the day, and specimens were eagerly sought by the leading naturalists. Wealthy English patrons such as Sir Joseph Banks sent their own collectors to the new colony. Banks financed the collectors and botanists George Caley, Allan Cunningham, Robert Brown and George Suttor and sponsored the two voyages of exploration of the Australian coast undertaken by Matthew Flinders. Until the death of Banks in 1820, the main scientific activity in the colony consisted in the collection of material for Banks and other British and European scientists. Ships sailed from Port Jackson loaded to the gunwales with plants and animals. James Hardy Vaux wrote in 1819 his ship was so crowded with kangaroos, emus and black swans that it resembled a Noah's Ark. Even aborigines, including the famous Bennelong, were 'collected' and sent to Britain for exhibition at scientific soirées.[2]

Just as the wool and cotton from her colonies went to the mills and factories of Britain for processing, so too went the colonial flora, fauna and fossils, to be processed in the scientific institutions and museums of Britain. British scientists were not interested in the opinions and theorizing of colonial naturalists and geologists. They simply imported the raw colonial data for their own theories and institutions. Indeed, Adrian Desmond has argued that the early colonial relations of Britain and

R.W. Home and S.G. Kohlstedt (eds.), International Science and National Scientific Identity, 55–74.

Australia, in accordance with which leading British naturalists and comparative anatomists viewed Australia as a 'faunal backwater, a kind of paleontological penal colony', were constitutive of their Eurocentric palaeontological and biological theories concerning Australian plant and animal life.[3]

At the very least, as this study demonstrates, the relations of colonialism structured the scientific activities of early naturalists as mere collectors for British theoreticians who were dependent upon the local collectors for the crucial raw material for their theorizing. Early colonial scientists were not only actively discouraged from theorizing but, lacking any scientific community of their own or colonial support for their science, they were in turn dependent upon British scientists and British institutions for scientific recognition and rewards.[4] British scientists, jealously guarding and promoting their own careers and reputations and their own institutional and social interests, were well placed to manipulate this colonial dependency to their own advantage, and to exert scientific hegemony over their colonial colleagues and colonial scientific specimens. It was this enforced emphasis on collecting, together with factors intrinsic to the structure of early colonial society and its remoteness from Europe, that shaped and in some cases retarded the growth of Australian science, technology and scientific institutions.[5]

This chapter will briefly examine the life-long relationship between the colonial naturalist Dr George Bennett and the eminent British palaeobiologist Sir Richard Owen as a case study of the British colonization of early Australian science. As I shall show, the talented and very competent local scientist could not achieve scientific recognition in Britain. Conversely, the London based curator could use Bennett's material and writings to enhance his own scientific reputation. Owen is the outstanding example of a British scientist who was able to become an expert on Australian materials – in his case Australian extinct mammals and Australian zoology – without ever setting foot on Australian soil. His formidable European reputation in this area was achieved and sustained by his widespread network of informants and collectors in Australia. Meanwhile Bennett, notwithstanding his own best efforts to achieve recognition in his own right, was unwittingly drawn into Owen's colonial network and into a subsidiary collecting role.

In the historical literature, Bennett is known only as a talented collector who sent unique specimens, fossils and living animals to the Royal Botanic Gardens at Kew, to the Zoological and Linnean Societies of

London, and especially to Richard Owen. Yet from his arrival to settle permanently in Australia in July 1835 until his death in 1893 at the age of eighty-nine, Bennett played a pivotal role in the development of Australian science. Upon his arrival, he was commissioned by the Governor, Sir Richard Bourke, to undertake a survey of the effects of influenza in sheep.[6] In the 1830s the colony's main income was from wool, and devastating diseases like catarrh or influenza were a threat to its economic future. Bennett's report, a milestone in the history of Australian science because it is one of the first government-commissioned scientific reports, indicates an early application of science to agriculture and marks the foundation of veterinary science in Australia.[7] As well, Bennett became a trustee of the Australian Museum, helped establish the Sydney School of Arts in 1838, and was a founder and active promoter of both the Acclimatisation Society (1861) and the New South Wales Zoological Society (1882). In 1856 Bennett, who was involved with science education, was appointed to the Board of Examiners of the recently founded University of Sydney, and for the last ten years of his life he was a co-examiner in Materia Medica and Therapeutics.

In his long lifetime, Bennett published over 115 papers, both in local and international journals. His interest in natural history was broad: he wrote on anthropological subjects such as cannibalism; on the cultivation of the orange; on fossils, ornithology and the reproduction of the kangaroo, platypus and the echidna; and on such medical matters as smallpox and tetanus and the incidence of infanticide amongst the Australian Aborigines. Bennett's two books, *Wanderings in New South Wales, Batavia, Pedir Coast, Singapore and China*[8] and *Gatherings of a Naturalist in Australasia*,[9] today are classics of nineteenth-century natural history literature. They are based on his own experiences and present a readable and informative account of native customs and the animals, birds, marine life and botany of the Asian and South Pacific region. Bennett's stated aim in *Gatherings of a Naturalist in Australasia* was to provide a 'truer picture' about the life, social condition and natural history of the Australasian region and so help to overcome the ignorance that existed in Britain concerning the colonies in the South Pacific.[10]

Bennett did receive some recognition for his collecting activities. In all, sixteen plants and animals were named after him. In naming the Queensland tree kangaroo (Dendrolagus bennettianus) after Bennett, Charles Walter De Vis, Director of the Queensland Museum, stated that he wished to 'dignify [the species] by association with one of our oldest and most respected Australian naturalists'.[11] Yet the real value of Bennett's scientific work has never been fully assessed. There are the brief, privately published biographical sketch by his devoted widow Sarah, V.M. Coppleson's 1955 oration, and Ann Moyal's publications, but there is no comprehensive biography that places Bennett within his social and political milieu, either in an Australian or a European context.[12]

In the past, Bennett has been too readily dismissed merely as Owen's 'collector'. Ronald Strahan, for instance, asserted that Bennett 'saw himself and the [Australian Museum] in a colonial setting, owing deference and service to the authorities at "home"'.[13] However, Strahan's assertion should be considered in the light of Bennett's own perception of himself as an Englishman working in a foreign land. In Australia in the 1830s, there was little sense of national identity, no concept of Australian citizenship and therefore no notion of scientific work relevant to an Australian nation. Bennett's letters reveal that he was not a 'mere' collector but rather an astute and active naturalist. He was purposeful and largely self-directed in his collecting, and he knew exactly what would interest the scientists at home in Britain. The value of his research, especially that on the reproductive systems of the kangaroo and the platypus, has been overshadowed or even forgotten by historians of science because of its appropriation by Owen for his own institutional and professional career purposes.

BENNETT-OWEN CONNECTION

Bennett's association with Owen, begun in London in the late 1820s when both were students at the Royal College of Surgeons, was an extremely one-sided affair. Bennett spent over fifty years collecting material for Owen, yet he received replies to only about one in ten of his letters.[14] Throughout the 1840s Bennett constantly chided Owen for his lack of acknowledgment of specimens sent to him and to the Hunterian Museum. Why did Bennett continue such an unrewarding association for so long? Donald Fleming's comment that colonial scientists maintained strong links with British scientists and associations 'in order to redress the psychological balance of the new'[15] may have some bearing on the problem. Ties with 'metropolitan' science reduced the sense of isolation of colonial scientists and encouraged them to believe that they were part of a larger global scientific community.

More to the point, when Bennett arrived to settle permanently in Australia in 1835, there was little in the way of a local scientific culture. Inevitably Bennett looked towards England for support and intellectual stimulation. Scientific societies and journals that began to be published in Sydney during the 1820s appeared only to flounder and fail due to a lack of money and a sufficiently numerous educated class interested in such things.[16] Apart from the Botanic Gardens and the Australian Museum, established in 1816 and 1827 respectively, no employment opportunities for scientists existed. Science in Sydney during the first half of the nineteenth century was very much in the hands of the wealthy amateur, and Bennett did not fit into this category – he had to work for his living. If he had had more money to pursue his bent for natural history, if there had been more professional opportunities in the colony of New South Wales, or if he had stayed in England, perhaps Bennett might have developed an international reputation. As it was, he was forced into a situation of colonial dependency on British science and its institutions and, above all, on his erstwhile fellow student, the ambitious and manipulative Owen, assistant curator of the Hunterian Museum at the Royal College of Surgeons and aspirant to the coveted title of the 'British Cuvier'.

I have selected for discussion three examples of Bennett's colonial discoveries that were used by Owen to greatly advance his own career: the Pearly Nautilus and the reproductive systems of the kangaroo and the platypus.

THE PEARLY NAUTILUS

The *Sophia*, a ship of 532 tons, having deposited its cargo of 92 male convicts in Sydney Town, was chartered by Captain Samuel P. Henry, not for a scientific expedition as stated by Coppleson but for a raid on the island of Erromanga (part of present day Vanuatu) to cut sandalwood. Sandalwood was then a very lucrative trade with the Chinese.[17] It was during this 1829 voyage of the *Sophia* that Bennett, then the ship's surgeon, made a most important scientific discovery, a living specimen of a Pearly Nautilus (Nautilus pompilius,Linn.) that he found floating in the waters of Dillon Bay on Erromanga. The Nautilus was an important find because it represented the sole living genus of the molluscan subclass Nautiloidea, which includes the fossil Ammonites.

Bennett's find was useful in relation to debate in England at the time on the age of the Earth. Here was a specimen that had links with the geological past – a living fossil. The soft parts of the Ammonite animals had never been found, but the close similarity of the Ammonite shells, bi-chambered and spiral in shape, to those of the Nautilus, implied that they had been occupied by similar squid-like animals. Bennett's Nautilus thus provided an analogue for the interpretation of the mode of life of the extinct Ammonites. The Pearly Nautilus and the even rarer Spirula, which Bennett was to find on a Sydney beach in 1860,[18] represented relics of spiral shelled species of cuttlefish that had swum in oceans of an ancient geological time.

The following extract from his journal reveals Bennett's skills as a scientific observer:

> Island of Erromanga, New Hebrides, August 24th, 1829. – Monday: fine weather during the day. Thermometer at noon 79 (degrees). In the evening a Pearly Nautilus (Nautilus Pompilius of Linnaeus) was seen in Marekini Bay, on the south-west of the island, floating on the surface of the water, not far distant from the ship, resembling, as the sailors expressed it, dead tortoise shell-cat in the water ... On its being brought on board, my attention was directed to possessing the inhabitant, which I succeeded in procuring. I immediately detached the animal from the fractured portions of the shell (to which it is attached by two oval muscular attachments, one on each side), and placed it in spirits, after making a pen-and-ink sketch of its external form. The animal when I examined it after it was brought on board, kept the tentacula closely contracted, and the only evidence of vitality remaining in the animal was in a slight contractile motion of the body. On laying carefully open that portion of the shell which contains the chambers, it was found to contain water, which of course immediately escaped ... The colour of the anterior part of the body, when the animal was recently taken out of the water, was of a dark red colour, inclining to brown ... The mantle and remainder of the body were of a light blueish tinge.[19]

On his return visit to England in 1831, Bennett published an illustrated account of his find in the *London Medical Gazette*.[20]

Over the years certain myths have grown up in reference to this discovery. One is that Bennett's specimen was sent personally to Owen. In fact a letter from Bennett dated 13 May 1831 states:

> Gentlemen
>
> I have the honor of presenting to the Museum of the Royal College of Surgeons the animal which inhabits the shell of the Nautilus Pompilius or Pearly Nautilus which I hope will be considered worthy of acceptance.[21]

Clearly Bennett, a recent graduate of the Royal College of Surgeons, was eager to achieve recognition from his peers as a naturalist/surgeon. It was not his intention simply to hand over his find to Owen.

However Owen saw immediately the scientific potential of Bennett's discovery and claimed the unique specimen for himself. With Bennett's Nautilus, Owen rushed to Baron Georges Cuvier for assistance. Cuvier, then head of the Musée d'Histoire Naturelle in Paris, was an inspiration and model to Owen. Even Owen's mother was aware of the value of his association with Cuvier. This is reflected in a letter to her son:

> Your being noticed by Cuvier was fortunate, and you having access to his museum would be an advantage in your profession on many accounts, and I trust you will reap the benefits of it ultimately.[22]

With the publication of his *Memoir on the Pearly Nautilus* in 1832, Owen achieved instant scientific fame. On the basis of Bennett's specimen, he modified Cuvier's classification of the Cephalopoda, proposing two orders of the species. These were considered valid until 1894.[23] This work also established Owen's reputation as a comparative anatomist. Even his arch-enemy, Thomas Huxley, writing in 1894, states that:

> the famous 'Memoir on the Pearly Nautilus ...' placed its author, at a bound, in the front rank of anatomical monographers. There is nothing better in the 'Mémoires sur les Mollusques', I would even venture to say nothing so good, were it not that Owen had Cuvier's great work for a model; certainly, in the sixty years that have elapsed since the publication of this remarkable monograph it has not been excelled.[24]

Above all, Owen used Bennett's Cephalopod to attack the theories of the famous French zoologist and physiologist Étienne Geoffroy Saint-Hilaire, who had attempted to link the Cephalopoda with the Vertebrata in demonstration of his morphological principle of unity of plan of all animal structures. As Toby Appel has shown, Geoffroy Saint-Hilaire aimed his morphology (which emphasized serial development, transmutation of composition, and the sovereignty of material laws) towards the young medical reformers and republicans of Paris, and against the conservative functional teleology and typology of Cuvier's comparative anatomy.[25] The structure of the Cephalopoda and how they should be classified was one of the central issues of the celebrated debate of 1830 between Geoffroy Saint-Hilaire and Cuvier, and was still a topical and contentious issue when Owen visited Cuvier in Paris in the following year. Predictably, the budding 'British Cuvier' sided with his chosen mentor.

In England, Owen's own conservative institution, the Royal College of Surgeons, was under heavy attack from the medical reformers who took their anatomical and ideological inspiration from Geoffroy Saint-Hilaire's British followers. According to Desmond's analysis, Owen aligned himself with the conservative institutional and social forces and structured his anatomy and palaeontology in opposition to the transmutationist assumptions of Geoffroy Saint-Hilaire and his followers, who assumed a serial development of life from monad to man.[26] Owen deployed Bennett's specimen in support of his anti-transmutationist assertion that the 'Cephalopods afforded one of the strongest arguments against the theory of simple and unbroken series' of life forms.[27] As Desmond has shown, Owen's conservative anti-transmutationist stance of the 1830s brought him rich professional and social rewards, along with the title of the 'British Cuvier'. Owen's appropriation of Bennett's Cephalopod must be understood in this context. Bennett's colonial claims stood little chance of taking precedence over Owen's own pressing institutional and social needs.

While Bennett maintained a lifetime interest in the Pearly Nautilus, writing on its ecology and utilization in the Pacific region, Owen lost interest in it with the publication of his *Memoir* and its gratifying scientific reception. It had served its purpose to Owen's Eurocentric concerns and, unlike Bennett, he had no interest in its indigenous context.

THE KANGAROO

The first known sighting and description by Europeans of a kangaroo or wallaby is thought to date back to 1629 when the Dutch merchant seaman, Francisco Pelsaert, was wrecked on the Houtman rocks off the coast of present-day Western Australia.[28] On inspecting these animals, Pelsaert described them as 'dogs'; and he expounded the theory that the young kangaroo or joey grew out from the nipple of the mammary gland located in the female pouch, thus establishing a myth about the reproductive system of the kangaroo that was to last for several hundred years.

The kangaroo was drawn to the attention of British scientists following the first voyage of James Cook to Australia in 1770. Both Cook and Joseph Banks described it as greyhound in size and running and having a tail as long as any greyhound.[29] On 4 August 1770 Cook wrote in his journal, 'the animal which I have mentioned is called by the natives "Kangooroo" "Kanguru"'.

An early biological account of the kangaroo and its young was published by Captain Watkin Tench. In 1788 Tench wrote:

Of the natural history of the kangaroo we are still very ignorant. We may, however, venture to pronounce this animal, a new species of opossum, the female being furnished with a bag, in which the young is contained; and in which the teats are found. These last are only two in number, a strong presumptive proof, had we no other evidence, that kangaroo bring forth rarely more than one at birth. But this is settled beyond a doubt, from than a doubt, from which more than a dozen females have been killed, which had invariably but one formed in the pouch.[30]

In the absence of foxes, kangaroo hunting became a common pastime in the colony of New South Wales. Both Darwin and Bennett were invited to participate in the sport. Despite the wholesale slaughter of these animals, which moved Bennett to issue a plea for their conservation,[31] even by the 1830s little was known about the system of reproduction.

On 4 February 1833, Bennett wrote to Owen of his discovery of a foetus of a kangaroo *in utero*. He sent Owen a bottle containing 'the uterus of a Kangaroo showing the foetus (with a placenta attached)',[32] together with a detailed sketch. Clearly aware of the significance of his find, Bennett again wrote to Owen on 20 February 1833:

My Dear Owen

The "Brothers" sailed last week for England having on board under the charge of my friend Dr. M'Ternan R.N. a box of preparation illustrating the generation of the Ornithorynchus and one showing the foetus of the kangaroo in utero, thus setting at rest the long pending question – I wish you to announce the discovery in my name together with wood carvings of the accompanying sketches of the preparation (which you can compare with it and oblige me by attending any inaccuracies) in any Journal or Naturalists' Magazine you may think proper – I leave it all to your judgement to do or leave alone as in your judgement you may think fit.[33]

Again Bennett was too trusting of Owen, for in the Royal Society of London's *Philosophical Transactions*,[34] it was Owen, not Bennett, who announced to the world the finding that young of the kangaroo developed not from the nipple in the pouch but within the uterus. In this paper, Owen acknowledged 'the exertions of my friend Mr. George Bennett'[35] but he failed to make reference to the above letter or to give credit to Bennet for the discovery. In obtaining the specimen, Owen appropriated the discovery and Bennett's field sketch of the uterus. As shown in Figures 1 and 2, even Owen's drawing of the uterus is similar, though admittedly Owen's shows the skill of more sophisticated draftsmanship – after all, it was drawn in the comfort of London. As Bennett had written in his letter of 4 February 1833, 'I had not the benefit of government men, bullocks, *et id omne genus*, and when 300 miles from Sydney, in the

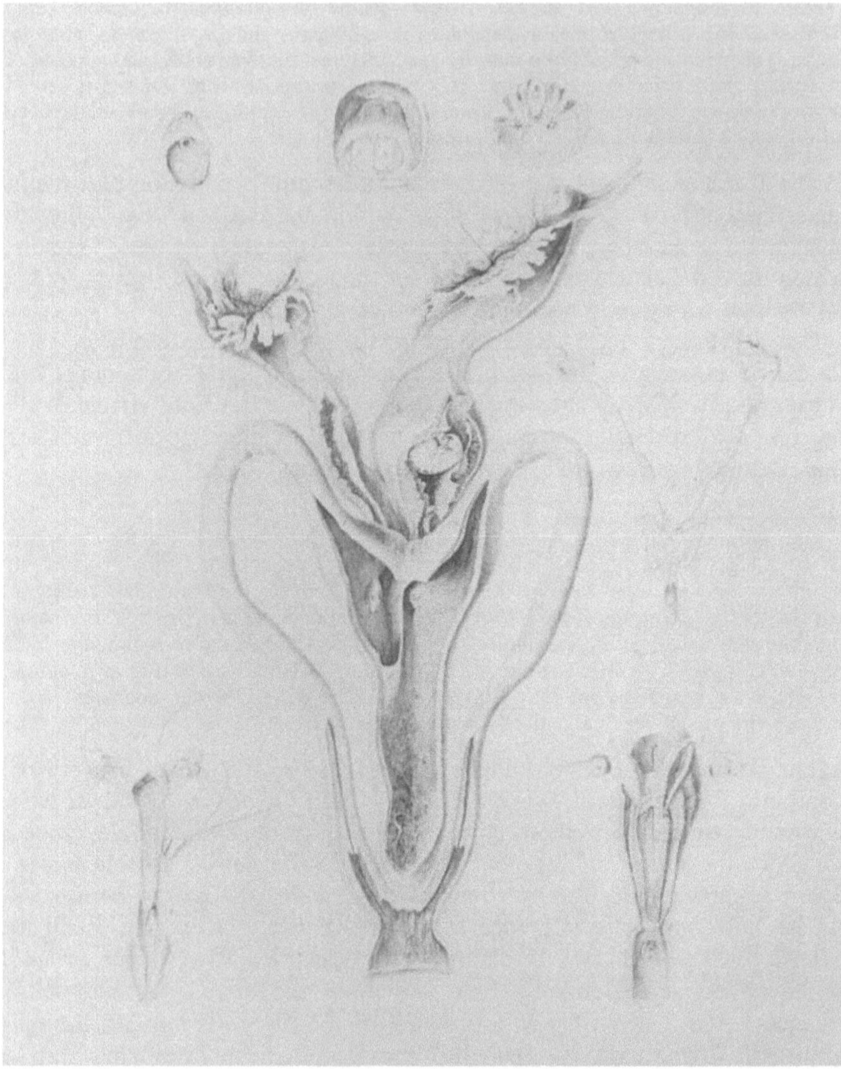

Fig. 1. Owen's drawing of the uterus of a kangaroo showing an embryo. (Source: 'On the Generation of the Marsupial Animals with a Description of the Impregnated Uterus of the Kangaroo'. *Philosophical Transactions*. (1834), 333–64.)

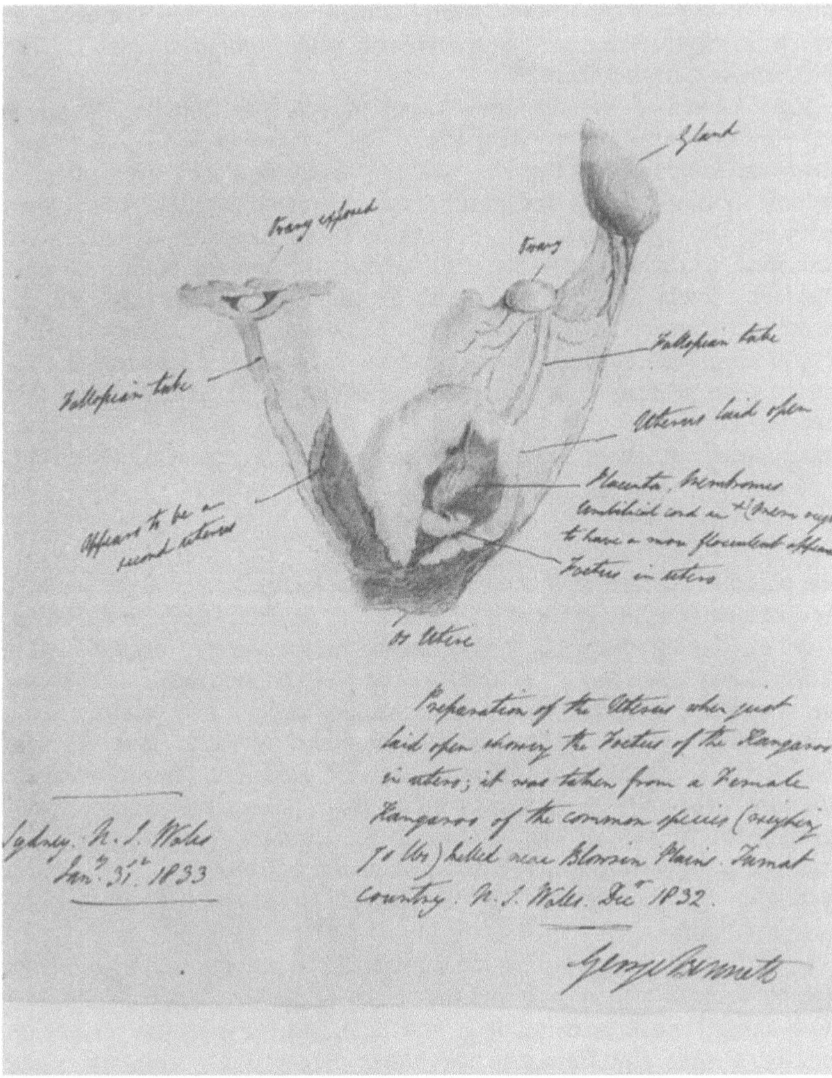

Fig. 2. Bennett's field sketch of the uterus and embryo of a kangaroo. (Source: Letter from Bennett to Owen, Sydney, 4 February 1833; Royal College of Surgeons, London, library).

interior on horseback, I found often difficulty in procuring something in which I might place my preparations, when procured, besides the difficulty of carrying them'.[36]

The problem of how the infant kangaroo was transferred to the pouch and nipple remained a mystery until 1964, when two CSIRO workers, observing kangaroos in captivity saw the young find their way, unaided, to their mother's pouch and attach themselves to the nipple, where they remained for eight months.[37] As early as 1806, American scientists had described a similar path for the embryo of another marsupial, the Opossum. Owen chose to ignore this finding and held the view that the infant was transferred to the nipple by the parent's lips, 'the mouth being always employed by the ordinary quadrupeds as Dogs, Cats and Mice for the purpose of removing their helpless offspring.'[38] Thus we see that theories formulated on the periphery, whether in Australia or the United States, were considered to be suspect and were not accepted by Owen.

THE PLATYPUS

The platypus (Ornithorhynchus) is unique to Australia, and, together with the Echidna or spiny anteater which inhabits Australia and New Guinea, is an egg-laying mammal. They are the sole survivors of the lowest mammalian class – the Monotremata. The word monotreme relates to the single aperture that functions in these animals as both an excretory and a reproductive system, a feature also common to birds and reptiles. Conclusive evidence of reptilian ancestry of the monotreme is its bone structure, especially the coracoid bones that connect the shoulder joint firmly with the rest of the skeleton at the sternum. This type of bone structure is only vestigial in other mammals, but has been an essential adaptation for the survival of the semi-amphibious platypus and the foraging echidna.

Apart from their egg-laying capacity, platypus possess true mammalian features such as a furry coat and the ability to suckle their young. Given these special characteristics, it is understandable that in the nineteenth century, British and European naturalists viewed these animals as the 'missing link' between reptiles and marsupials. Throughout the century, naturalists grappled to find the answer to the taunting questions, were these animals oviparous, ovoviviparous or viviparous, and did they suckle their young?

The first recorded sighting of the platypus by a European was made on the banks of the Hawkesbury River, north of Sydney, by David Collins in November 1797.[39] The first printed account, by Dr George Shaw of the British Museum, appeared in the tenth volume of the *Naturalist's Miscellany* in 1799. On its arrival in Britain and Europe, no other animal was so controversial to both layman and scientist. It was initially regarded as a hoax. As Robert Knox wrote in 1823:

> Since these animals reached England by vessels which had navigated the Indian seas, a circumstance in itself to rouse the suspicions of the scientific naturalist, aware of the monstrous impostures which the artful Chinese had so frequently practised on European adventurers; in short the scientific felt inclined to class this rare production of nature with eastern mermaids.[40]

This secretive mammal created a long standing mystery in scientific debates on evolution during the 1830s and '40s. Literally thousands of preserved and stuffed platypus and their uteri were shipped back to Britain for research. It is a miracle that they still survive today after such slaughter. As early as 1819, the judge/scientist Barron Field, working in Australia, was moved to express concern for the conservation of the platypus (or water mole as it was known locally). In one of his poems he wrote of a time 'when sooty swans are once more rare, And duck-moles the Museum's care'.[41] Sums of money were offered to solve the riddle, but without success. Sir Everard Home in 1802 even suggested that the Colonial Secretary be approached to sponsor research into the life of the platypus.[42] The debate over the reproduction and classification of the platypus was part of a larger debate between the pro- and anti-transmutationists of the 1830s and '40s, and inevitably, it engaged Owen's attention in his anti-transmutation and anti-Geoffroy Saint-Hilaire crusade. In the 1830s, the French anatomist had classified the 'Monotremata' (he coined the term) in a separate grouping between the reptiles and the mammals. Their transitional status was clearly of significance to Geoffroy Saint Hilaire's transmutationism, and he justified his categorization of the platypus as non-mammalian on the grounds of its supposed oviparous mode of reproduction and its lack of mammary glands. Owen, in opposition to Geoffroy Saint Hilaire, insisted on the mammalian nature of the platypus and argued that it was ovoviviparous. In this dispute, by virtue of the British monopoly over colonial platypus specimens, Owen had the advantage of almost exclusive access to the essential material for dissection. But while Owen made the most of the available preserved specimens in support of his position, the dispute could only be resolved by first-hand reports from the field. Bennett was well qualified for the job. Not only was he a surgeon, he had trained under Caesar Hawkins, the first surgeon in London to perform an ovariotomy. During his visit to England in 1832, Bennett perceived the scientific significance of the platypus and resolved to tackle the contentious issue of its mode of reproduction. He feared that he might be pre-empted before he could return to Australia.[43]

Upon his arrival in Sydney in August 1832, Bennett lost no time in the pursuit of the elusive platypus. He immediately embarked on the arduous crossing of the Blue Mountains, even though the roadway was still incomplete and the weather freezing.

His most important find on this journey was that the platypus did suckle its young, and he confirmed J.F. Meckel's 1824 description of the animal's mammary glands. These glands are very different from those of other mammals, being actually modified sweat glands; that is, there is no teat but rather minute pores that exude milk. During the breeding season the glands become engorged but they become very reduced after this period. Consequently the mammary glands in this animal had been overlooked, and this had led to Geoffroy Saint-Hilaire's conclusion that the platypus was not a mammal.

In his letter to Owen dated 4 February 1833, Bennett also wrote of this discovery:

> I have a section of one female Ornithorynchus I shot, in which the milk gland is very large and I can now inform you from actual observation that milk is secreted from it, it comes out (as your mercury did when you injected the ducts) in small drops on the surface of the skin ... you can mention it to the Zoological Society as a decided fact; and which has also been seen by some intelligent gentlemen in this Country – but I was not satisfied, to assert it, until I became an eye-witness of the fact.[44]

Bennett was so excited by his find that he requested Owen to show his specimens to the Zoological Society with 'some brief comments in my name'. Subsequently an account of Bennett's findings was communicated by him to the Zoological Society on 27 May 1834. On the same day, Owen communicated his own findings, based on an examination of specimens received from Bennett. Curiously Owen, whilst giving praise to Bennett for a speedy elucidation of the mammary gland question, used results, based on the examination of the contents of the stomach of a young platypus sent by Bennett, to convey the impression that he, Owen, had personally discovered that the platypus produces milk for its young.

In the same letter, Bennett reported another and, as far as Owen was concerned, even more important and alarming find, namely that the monotremes were oviparous:

I have much pleasure in sending you by the return of the ship "Brothers" some preparations ... which no doubt you will consider valuable; there are some proving Ornithorynchus oviparous.[45]

Owen would not have supported an oviparous stance, for it would have meant agreeing with his French rival Geoffroy Saint-Hilaire. In 1832 he had dismissed the find of Lieutenant Lauderale Maule of the 39th Regiment, who had found several nests of the Ornithorynchus in which, whilst there were no whole eggs, there were pieces of 'substance resembling egg-shells'.[46] On that occasion Owen had explained away these apparent egg shells as the urine excrement of the platypus and inconclusive proof of the oviparous nature of the platypus.[47]

Maule had been attempting to prove the persistent rumours amongst the colonists that the platypus laid eggs. Unfortunately these rumours were ignored as unscientific by the British and European mammalian experts. Bennett had sent to Owen a specimen of a female platypus containing an egg in the left uterus. However, whilst Bennett at this stage clearly believed that the platypus laid eggs, since he had found no nests for the hatching of these eggs, he could not be certain that the animal was indeed oviparous. It was still possible that the animal was ovoviviparous, that is, that the eggs were hatched inside the mother and the young were born alive. This was the opinion held by Owen. Many field workers such as Bennett enlisted the aid of Aborigines, without success. In Bennett's case he seems to have become confused by their statement that 'plenty pickaninny tumble down from old women'.[48] In any case, under Owen's influence he changed from believing these animals were oviparous to an ovoviviparous stance.

Another problem associated with researching the reproductive system of the platypus is that, like birds, they have a limited breeding season – depending on the geographical region, from August to October. Since the hatching period is short, according to Burrell only three weeks,[49] the field worker had to be there at the right time. Another problem was that up until the 1920s it proved impossible to breed these animals in captivity. Bennett, an experimentalist, made numerous attempts but failed.[50]

In 1884, William Hay Caldwell, a Balfour Scholar from Cambridge, announced to the world: 'Monotremes oviparous, ovum meroblastic'. The mystery of the monotremes had been solved. Graciously, Bennett accepted Caldwell's find:

Who would have thought that an animal with no large milk gland should actually demean itself by laying small white eggs, that it was oviparous in the uterus we knew in 1834 but I had come to the conclusion that it would be ovi-viviparous. I shall look forward with the greatest interest to see the embryology of this singular animal well worked out by Mr. Caldwell.[51]

It would seem that although Bennett had, on the basis of his own researches and local information, been certain of the oviparous nature of the platypus in 1834, he subsequently succumbed to the nationalistic pressures of 'Empire'. Bennett could not sustain his 'pro-French' oviparous stance in the contemporary context of the debate between the British Owen and the French Geoffroy Saint-Hilaire. In any case, the mammalian nature of the Monotremes was established by Bennett's observation of milk secretion from the mammary glands and his specimens of female platypus at different stages of the sexual cycle. With his appropriation and interpretation of this colonial material, Owen was able to close the debate in favour of British science and his own institutional and social interests.

CONCLUSION

The three examples discussed in this paper demonstrate the value of Bennett's scientific skills to the advancement of Owen's professional career and to the consolidation of British comparative anatomy and palaeontology. From his central scientific and institutional vantage point, Owen appropriated Bennett's specimens and observations. Bennett, working at the periphery of the Empire and its science, had far fewer opportunities to cultivate power and prestige, and was easily marginalized and subordinated by the ambitious and unscrupulous Owen. Bennett was not the only scientist to be colonised by Owen. Jacob Gruber has shown recently [52] how Owen in England became the centre of a 'Moa cottage industry'. New Zealand provided him with a continuous flow of specimens that he 'processed' for his own professional advancement and that of an expanding community of British professionals. Owen did very nicely out of the New Zealand moa bones. Sir Richard Peel awarded him a Civil List Pension of £ 200 per annum largely for his reconstruction of the moa.

Although his letters show that Bennett (initially, at any rate) perceived himself as Owen's peer and expected his help and support in communicating his discoveries, it seems clear that if Bennett wanted to gain any scientific recognition in Britain, he had little choice but to submit to the subordinate collecting role assigned to him by Owen. Colonial relations structured scientific relations. Given his lack of a local scientific community and institutions, Bennett was powerless to resist the expropriation of his discoveries and their deployment by Owen for his own institutional and social purposes. Bennett wrote poignantly that at times he felt 'scientifically alone in Australia'.[53] It is notable that the young Charles Darwin and Thomas Huxley fared much better than Bennett at Owen's hands. They were on the spot in London, not 12,000 miles away in Sydney.

Symbolically, the College of Surgeons awarded Bennett an honorary Gold Medal in recognition of the value of his shipments, but it was Owen who was made a Fellow of the Royal Society for his Pearly Nautilus and platypus work. Desmond has asserted that Owen's success 'testified to the superiority and exclusivity of his material, which in turn reflected continuing colonial expansion in New South Wales'.[54] But really Owen's scientific success reflected the successful British colonization and domination of early Australian science.

University of Wollongong
Australia

NOTES

* I should like to thank Dr Evelleen Richards for her assistance and patience in the revision of this paper and the British Museum (Natural History), the Royal College of Surgeons of London, the Mitchell Library and the State Library of New South Wales for the opportunity to study and quote from manuscript material in their keeping.

1 T.H. Huxley to William Sharp Macleay, 9 November 1851, quoted in Leonard Huxley, *Life and Times of Thomas Huxley* (London, 1900), p. 257.

2 *Memoirs of James Hardy Vaux* (London, 1819), p. 205; quoted in Geoffrey Serle, *From Deserts the Prophets Come: the Creative Spirit in Australia, 1788–1972* (Melbourne, 1973), p. 4.

3 Adrian Desmond, *Archetypes and Ancestors: Palaeontology in Victorian London, 1850–1875* (Chicago, 1982), p. 143.

4 M.E. Hoare, 'Dr. John Henderson and the Van Diemen's Land Scientific Society', *Records of the Australian Academy of Science*, 1 (3) (1968), 7–24; pp. 20–21.

[5] Elizabeth Newland, 'Sir Roderick Murchison and Australia: a Case Study of British Influence on Australian Geological Science', (MA thesis, University of New South Wales, 1983), pp. 19–57.
[6] *Report on the Epidemic Catarrh or Influenza Prevailing among the Sheep in the Colony of New South Wales in the Year 1835*, by George Bennett, Esq, Andrew Gibson, Esq. J.P. and William Sherwin, Esq. (Sydney, 1835).
[7] Geoff Raby, 'Science in the "Pastoral Age": Veterinary Responses to the New South Wales Catarrh Epizootics of the 1830s', *Historical Records of Australian Science*, 6 (2), (1985), 189–94, p. 193.
[8] George Bennett, *Wanderings in New South Wales, Batavia, Pedir Coast, Singapore and China: being the Journal of a Naturalist in Those Countries, during 1832, 1833 and 1834* (London, 1834), 2 vols.
[9] Bennett, *Gatherings of a Naturalist in Australasia: being Observations principally on the Animal and Vegetable Production of New South Wales, New Zealand and some of the Austral Islands* (London, 1860).
[10] Ibid., Preface.
[11] V.M. Coppleson, 'The Life and Times of Dr. George Bennett: Annual Post-Graduate Oration', *Bulletin of the Post Graduate Committee in Medicine, University of Sydney*, 11 (1955), 206–64; p. 229.
[12] *George Bennett, MD, FLS, FZS, etc.: Biographical Sketch* (Sydney, n.d.); Coppleson, op. cit. (n. 11); Ann Mozley Moyal, *Scientists in Nineteenth Australia: a Documentary History* (Melbourne, 1976); idem, 'Richard Owen and His Influence on Australian Zoological and Palaeontological Science', *Records of the Australian Academy of Science*, 3 (2) (1975), 41–56; idem, '*A Bright and Savage Land*': *Scientists in Colonial Australia* (Sydney, 1986).
[13] Ronald Strahan, *Rare and Curious Specimens: an Illustrated History of the Australian Museum, 1827–1979* (Sydney, 1979), p. 17.
[14] Coppleson, op. cit. (n. 11), p. 216.
[15] Donald Fleming, 'Science in Australia, Canada and the United States: Some Comparative Remarks', *Proceedings of the 10th International Congress of the History of Science, Ithaca 1962*, p. 182.
[16] Elizabeth Newland, 'Forgotten Early Australian Journals of Science and Their Authors', *Journal of the Royal Australian Historical Society*, 75 (1986), 59–68.
[17] Coppleson, op. cit. (n. 11), p. 209. Dorothy Shineberg, in her book *They Came for Sandalwood: A Study of the Sandalwood Trade in the South-West Pacific, 1830–1865* (Melbourne, 1967), provides a detailed account of Bennett's and the *Sophia*'s voyage to Hawaii, Tahiti, Tonga, Fiji and the New Hebrides seeking sandalwood for the China trade.
[18] W.J. Dakin, *Australian Seashores* (Sydney, rev.ed., 1960), p. 306.
[19] Bennett reported his find in *Gatherings of a Naturalist in Australasia* (op. cit. [n. 9]), p. 377. An account also appears in Richard Owen, *Memoir on the Pearly Nautilus (Nautilus pompilius, Lin.)* (London, 1832).
[20] Bennett, 'Pearly Nautilus', *London Medical Gazette*, 8 (1831), pp. 729–30.
[21] Letter from Bennett to Royal College of Surgeons, London, 13 May 1831; Library of the Royal College of Surgeons, London.
[22] Rev. Richard Owen, *The Life of Richard Owen* (London, 1894), p. 57.

[23] See 'Owen, Richard', *Dictionary of Scientific Biography*, vol. 10, pp. 260–3.
[24] Rev. Richard Owen, op. cit. (n. 22), p. 306.
[25] Toby A. Appel, *The Cuvier-Geoffroy Debate: French Biology in the Decades before Darwin* (Oxford, 1987).
[26] Adrian Desmond, 'Richard Owen's Reaction to Transmutation in the 1830s', *British Journal for the History of Science*, 18 (1985), 25–50.
[27] Owen, op. cit. (n. 19), p. 1.
[28] Peter Stanbury and Graeme Phipps, *Australian Animals Discovered* (Sydney, 1980), p. 31.
[29] Ibid.
[30] Ibid., pp. 32–33.
[31] Bennett, op. cit. (n. 8), p. 278.
[32] Letter from Bennett to Owen, Sydney, 4 February 1833; Library of the Royal College of Surgeons, London.
[33] Letter from Bennett to Owen, Sydney, 20 February 1833; Library of the Royal College of Surgeons, London.
[34] Owen, 'On the Generation of the Marsupial Animals, with a Description of the Impregnated Uterus of the Kangaroo', *Philosophical Transactions*, 125 (1834), 333–64.
[35] Ibid., p. 335.
[36] Bennett, op. cit. (n. 33), p. 1.
[37] G.B. Sharman and J.H. Calaby, 'Reproductive Behaviour in the Red Kangaroo, Megaleia rufa, in Captivity', *CSIRO Wildlife Research*, 9(1964), 58–85.
[38] Owen, op. cit. (n. 34), p. 345.
[39] Stanbury and Phipps, op. cit. (n. 28), p. 13.
[40] Robert Knox (1823), quoted in Harry Burrell, *The Platypus: Its Discovery, Zoological Position, Form and Characteristics, Habits, Life History* (Sydney, 1927), pp. 1–2.
[41] Ibid., p. 193
[42] Ibid., p. 36.
[43] Bennett, 'Notes on the Natural History and Habits of the Ornithorhynchus paradoxus. Blum.', *Transactions of the Zoological Society of London*, 1(1835) 229–258; p. 229.
[44] Bennett, op. cit. (n. 32), p. 7.
[45] Ibid., p. 1.
[46] Maule's discovery was reported by Owen in his paper, 'On the Mammary Glands of the Ornithorynchus Paradox', *Philosophical Transactions*, 123(1832), 517–38; pp. 53–4.
[47] Ibid., p. 534.
[48] Bennett, op. cit. (n. 43), p. 240.
[49] Burrell, op. cit. (n. 40), pp. 171–2.
[50] Accounts of his attempts at raising these animals are published in Bennett, op. cit (n. 9), pp. 133–9, and Bennett, op. cit. (n. 43), pp. 257–8.
[51] Letter from Bennett to Owen, Sydney, 23 December 1844; British Museum (Natural History) Library.
[52] Jacob W. Gruber, 'From Myth to Reality: the Case of the Moa', *Archives of*

Natural History, 14(1987), 353–65.
[53] Letter from Bennett to Owen, Sydney, 10 March 1836; Library of the Royal College of Surgeons, London.
[54] Desmond, op. cit. (n. 26), p. 40.

ROBERT A. STAFFORD

A FAR FRONTIER:

British Geological Research in Australia During the Nineteenth Century

Geology in Australia, from the end of the eighteenth century until at least the First World War, remained primarily the preserve of British geologists. Given Australia's colonial status during the greater part of this period, it was to be expected that emigrant scientists from the metropolis would colonize employment opportunities arising at the periphery of the imperial system. Britain's tradition of amateur scientific inquiry also dictated that a good deal of such research would be accomplished by men who were not full-time scientists. As geology was professionalized and institutionalized in Britain during the second half of the nineteenth century, Australian geology likewise became the province of the geological surveyor or professor rather than the avocation of the military officer or clergyman. Yet long after the foundation of Australian universities and schools of mines, these professional posts continued to be filled almost exclusively by Britons.

In many respects, Australian geology in the nineteenth century was British geology done at a remove. Wade Chambers has provided a corrective to the myth that Australian science has been tyrannized by distance. Geologists working in the Australian colonies were nevertheless isolated, hampered by alien field conditions, hindered by the lack of topographical base maps, bedevilled by a philistine cultural milieu, and frequently confused by unfamiliar fossils and formations. Still, the discipline was practised in Australia very much as it was in Great Britain and Ireland. Scientifically, as well as economically and culturally, the colonies represented distant provinces of the mother country. The geologist Joseph Jukes expressed this relationship in organic imagery: 'An Englishman knows nothing of the power and grandeur of his country who has remained at home all his life. England is but the heart of the empire. He only knows the stature of the whole body who has seen the limbs embracing the whole world, and peoples, kingdoms, and nations reposing under the shadow of its arm.'[1] As the continent comprising Britain's remotest colonies, Australia constituted the farthest frontier of

R.W. Home and S.G. Kohlstedt (eds.), International Science and National Scientific Identity, 75–96.

nineteenth-century British geology. This essay will examine the means by which the discipline was exported to the Antipodes, its accomplishments there, and the relationships that linked the Australian and metropolitan branches of the science at this period.

In 1830, at the height of the 'heroic' age of British geology, Charles Lyell offered three pieces of advice to the beginning geologist: 'Travel-travel-travel'.[2] Roderick Murchison, who encouraged exploration throughout the world in order to expand the scope of his own Silurian, Devonian, and Permian stratigraphic systems, similarly asserted 'the incontrovertible axiom that physical geography and geology are inseparable scientific twins'.[3] Geology thus fed and grew upon new data. As the developing discipline gained its first specialized institution in 1807 with the founding of the Geological Society of London, British geologists set out to build up an accurate corpus of empirical data as a prerequisite to useful theorizing about the evolution of the Earth.[4] They concentrated their research on the stratigraphic enterprise – the chronological ordering of the layered rocks of the Earth's crust and the mapping of their surface outcrops.

As a result of awareness of the economic advantages to be won from agricultural improvement, mining and quarrying, many Fellows of the Geological Society took an interest in the practical applications of their discipline. Most geologists were members of the landed or commercial élites, or of the professional classes that served their interests. Such men pursued the science for its intellectual rewards and social cachet, but they were also alive to the potential role of stratigraphy in guiding mineral exploitation.[5] While the élites and their ancillary 'service class' shared a desire to apply geology to the improvement of property revenues in Britain, they demonstrated a like interest in using the science to increase profits from the vast undeveloped estates represented by India and the colonies.

The activities of Sir Joseph Banks nicely illustrate this fusion of purpose. His family owed its fortune to estate improvements and mining ventures. Banks himself, as an enthusiast for imperial development, remained primarily interested in botanical projects aimed at creating new colonial agricultural industries. But he also employed agents to collect specimens of valuable minerals throughout the empire, provided geological instructions for naturalists on Admiralty exploring expeditions, and arranged that a miner assist the botanist Robert Brown in his search for exploitable minerals during Matthew Flinders's voyage to chart the coasts of Australia.[6] Banks's friend and rival Charles Greville likewise combined an interest in estate improvement, mining investment and port development with an enthusiasm for creating a mineralogical collection of imperial scope that might serve as a basis for systematic resource development.[7] The Royal Institution in London, where Greville planned to house his collection, also exemplified the meshing of interest in science as an instrument of domestic and imperial development.[8] Indeed, the potent alliance between upper- and middle-class interests that was formed to achieve the economic and social goals embodied in 'gentlemanly capitalism' served as the principal engine of British overseas expansion from the seventeenth to the mid-twentieth centuries.[9]

With the ending of the wars with France, British geologists began to look abroad for fresh fields in which to test the validity of the stratigraphic divisions they were defining. Continental geologists played a key role in the construction of an internationally accepted stratigraphic column, but British geologists led this field of research. Besides being the workshop of the world, Britain emerged in the early nineteenth century as a storehouse of typical stratigraphic sections. The export of Britain's system of stratigraphic nomenclature was consciously paralleled by her geologists to the export of her manufactured goods. The emphasis on mapping also gave geology a strong territorial dimension, so that the vast tracts overseas coloured according to the British system represented new provinces annexed to the domain of British science. Beyond Ireland and the Continent, the colonies beckoned with a diverse range of unexploited scientific frontiers. It was clear that the ability of geology to discover new sources of revenue, to further plans for economic development, and to contribute to the practical maintenance of British rule might be well appreciated by colonists and administrators alike.[10]

In Australia, practical concerns such as the provision of building materials dominated interest in geological matters during the first years of settlement. Discoveries of coal and iron, however, coupled with the influence of Greville in London, led to the appointment of the mineralogist Adolarius Humphrey to the civil establishment of New South Wales in 1803.[11] Humphrey accomplished very little before his dismissal in 1812, but other naturalists and explorers such as the botanist Allan Cunningham performed a good deal of geological research in subsequent years. From 1823 to 1837 John Busby served as Mineral Surveyor in New South Wales, but his energies were chiefly absorbed by his other duties as the colony's Civil Engineer. Geological research in

Australia thus began under the patronage of the imperial government, and utilitarian exigencies rather than scientific curiosity were its motive force.

In the United Kingdom, a few geologists were patronized by the government on a temporary basis during the same period. This pattern culminated in Ireland with the establishment of a Geological Branch of the Board of Ordnance in 1826 and in England with the hiring of Henry De la Beche in 1832 to colour geologically the topographical maps being produced by the Ordnance Survey. De la Beche, a Fellow of the Geological Society whose leisure for research had become threatened by the failure of his Jamaican sugar estate, thus convinced the government to create a post that permitted him to pursue his scientific interests in the style appropriate to a gentleman. His tenure was extended in 1835 by the creation of the Geological Survey of England and Wales, the first national geological survey in the world. By 1845, this establishment had been reconstituted as the Geological Survey of Great Britain and Ireland, De la Beche had been knighted and given the title of Director-General, a Mining Record Office and museum had been added, and laboratory specialists representing several disciplines had been hired to complement the expanding field staff.

In 1851, when the Museum of Practical Geology was opened in London, a School of Mines (later designated the Royal School of Mines) was established at the same location. The Museum displayed fossils and specimens of economic minerals collected throughout the British empire, as well as models of mining technology. The School was intended to produce recruits for the parent Survey as well as mining engineers and geologists to initiate a new era in efficient mining. It was also envisaged as a nucleus for a national system of scientific education aimed at increasing the competitiveness of British industry.[12]

From the start, De la Beche emphasized the commercial and administrative utility of his Survey. The imperial sphere was an obvious arena for such demonstrations, and De la Beche read well the centrifugal trajectory of gentlemanly capitalism. During the 1840s, he used his laboratories to evaluate coal supplies throughout the world on behalf of the Royal Navy. He also recommended geologists for Canada and India, and the colonial surveys thus founded were modelled on the parent establishment. De la Beche, while beating the drum of practicality, nevertheless managed to direct an ambitious programme of research regarding the relationship between organisms and their physical environments during past ages.[13] In an era of laissez-faire, colonial geological surveyors were constrained by a need to demonstrate the usefulness of their activities in order to justify their funding. The primary role of official geology in the imperial context was thus exploratory – to call forth latent productivity from under-developed colonies by discovering mineral resources and providing information to guide their exploitation.

Throughout the empire, geological inquiry proceeded desultorily during the first half of the nineteenth century. In Australia, French naval exploring expeditions in the late eighteenth century had accomplished a good deal of coastal geological observation, but the defeat of Napoleon put paid to any French hopes of teritorial acquisition in Australia and thus of systematic inland surveys. The geologists of those European powers possessing overseas empires enjoyed definite advantages over their colleagues in nations not so endowed, for while distant colonies were not a prerequisite of large-scale field research and disciplinary advancement, they nevertheless helped. Empires offered extended scientific frontiers with opportunities for subsidized research and long-range stratigraphic correlation.

These opportunities were circumscribed by government policy and social reality in the penal colony of Australia. Exploration of the interior was slow in coming. Hydrographical expeditions continued to function as a major vector for geological research until nearly mid-century. Phillip Parker King's coastal surveys, carried out between 1817 and 1822, yielded observations and specimens that when analysed in London helped sketch in the geology of much of the coast.[14] Charles Darwin undertook excursions in Tasmania and New South Wales during the cruise of the *Beagle* in 1836, making observations of the Blue Mountains gorges and collecting fossils that helped elucidate the Australian Palaeozoic succession.[15] Between 1842 and 1846, J.B. Jukes twice circumnavigated Australia as naturalist on the H.M.S. *Fly* hydrographical expedition. Jukes had studied at Cambridge under Adam Sedgwick, founder of the Cambrian stratigraphic system, and brought the experience gained during an official survey of Newfoundland to his task in the Antipodes. Jukes too made valuable contributions to knowledge of the Palaeozoic succession in eastern Australia, and he was the first trained geologist to examine the continent's northern coastline. The account that he published upon his return to Britain and appointment to the British Survey constituted the first coherent overview of the geology of all Australia.[16]

Meanwhile geological investigations in the colonies themselves continued. Thomas Mitchell, Surveyor-General of New South Wales and a Fellow of the Geological Society of London, collected fossils on his exploring expeditions during the 1830s that, when classified in London, greatly extended knowledge of Australian stratigraphy. Mitchell also prepared Australia's first published geological map and conducted the first scientific cave explorations in the continent, discovering a spectacular series of fossils of giant extinct marsupials that bore importantly upon the European debate regarding the progressive development of life.[17] Charles Sturt's explorations of the Murray River and the hinterland north of Adelaide likewise supplied English savants with field evidence from the interior.

In Tasmania, Lieutenant-Governor Sir John Franklin, a Fellow of the Royal and Geological Societies, patronized science by founding a natural history society.[18] This represented an Antipodean version of Britain's provincial literary and philosophical societies, and was of a piece with similar institutions being established in India, South Africa and Canada. The van Diemen's Land Society, which published a well respected journal, acted as a magnet for colonists interested in science and for visiting savants. Franklin himself patronized the geological investigations of Johann Lhotsky, a Czech naturalist collecting specimens in Australia for European museums. Franklin provided Lhotsky with an introduction to Murchison, to whom the Czech showed his specimens in 1838. Judging from this and other evidence that Palaeozoic rocks might prove widespread in southern Australia, Murchison requested more fossils from Franklin. The governor then enlisted the aid of Paul Edmund de Strzelecki, a Polish explorer who also displayed his specimens to Murchison as well as the manuscript and map of his *Physical Description of New South Wales and Van Diemen's Land*. This evidence convinced Murchison that his Silurian system was well developed in Australia.[19]

A complex network of interchange thus linked Australian and British geology at both the individual and institutional levels. Other national influences were operative as well – Lhotsky and Strzelecki exemplify Continental interest in Australian geology, and French naval explorers continued to provide specimens for French palaeontologists and stratigraphers. In 1839 the American geologist James Dwight Dana also visited Australia with the United States Exploring Expedition. Dana correctly postulated a fluvial origin for the valleys of the Blue Mountains and eventually concluded that the coal measures of New South Wales were late Palaeozoic, but younger than the major coal deposits of the northern hemisphere.

A more significant involvement in Australian geology emerged on the part of Germans. Johannes Menge served briefly as geologist to the South Australia Company following initial settlement of that colony in 1836, and Dr Ferdinand von Sommer conducted mineral explorations for the Western Australian government in the late 1840s.[20] The Prussian, Ludwig Leichhardt, also made valuable stratigraphic observations during his explorations. Many German scientists, having imbibed liberal political sentiments at university, fell foul of reactionary German authorities and fled abroad to preserve their liberty. Britain constituted the chief haven for such political refugees, and several forged careers in the colonies. Lhotsky as well as Ernst Dieffenbach, naturalist to the New Zealand Company from 1839 to 1842, exemplify the influence of this political factor upon science in the Antipodes.[21] Lacking colonies or a high seas fleet of her own, pre-unification Germany offered limited opportunities for foreign travel, even to scientists not alienated by the repressive political climate. For them, too, Britain beckoned with its global empire, omnipresent navy, and German Prince Consort eager to patronize the researches of his countrymen. A good deal of German scientific work was thus performed on British imperial service during the Victorian era, and Australia was a major locus for such activity.[22]

Inevitably, however, British scientists remained pre-eminent in the Antipodean field. Australia – and, after its annexation in 1840, New Zealand – remained exclusive British preserves. Despite the rhetoric of internationalism and the willingness of British scientists and administrators to employ the results of foreign sojourners in the colonies for their own ends, underlying rivalries and cultural assumptions dictated that British geologists would view their work as serving imperial interests.

The most important figure in Australian geology before mid-century was the Anglican clergyman William Branwhite Clarke, who emigrated to New South Wales in 1839. Like Jukes, Clarke had trained at Cambridge with Sedgwick. He came out to Sydney as a Fellow of the Geological Society with an established scientific reputation, and he quickly became an important member of the small scientific circle centered on the Australian Museum. While conducting his own programme of fieldwork and collaborating with visiting scientists such as Dana and Jukes, Clarke's orientation remained British. He devoted much attention to establishing the stratigraphic age of the New South Wales coal deposits. He also discovered gold in the Blue Mountains in 1841, and after further finds announced the fact to Governor Gipps in 1844.

Clarke was not the first to find gold in Australia. Strzelecki, for example, discovered traces of it in 1839 and his report was published as a British parliamentary paper. His reference to gold was obscure, however, and it was years before either he or Clarke received recognition for establishing the metal's existence on geological grounds. The delay was due to unofficial censorship by Gipps, who feared the results of such news during the colony's transition from penal to free society. Yet Clarke began campaigning for the establishment of a geological survey in New South Wales, and after Gipps' death in 1847 he began publicizing the probable abundance of gold in the colony. Jukes and Strzelecki had also urged the foundation of such a survey. Murchison, who after examining Strzelecki's material had in 1844 predicted the discovery of gold in Australia on the basis of a comparison of the Great Dividing Range and the auriferous Ural Mountains of Russia, had likewise importuned the Colonial Office. Yet nothing was done to answer these demands until 1849, when the accelerating pace of mineral discoveries finally brought about the establishment of a geological survey in New South Wales.[23]

The geologist appointed to this post at De la Beche's recommendation was Samuel Stutchbury, an English museum curator and coal viewer who had visited Sydney in the 1820s as a commercial naturalist.[24] Jukes as well as Henry Bristow, another member of the home Survey, had declined the post prior to Stutchbury's acceptance, for though the salary of £ 600 per annum would have doubled their incomes, the appointment was envisaged as temporary. Stutchbury was paid by the government of New South Wales, but in scientific terms he remained an agent of the British geological establishment. De la Beche stipulated that Stutchbury conduct his research in conformity with the methods of the home Survey and that he collect specimens for the Museum of Practical Geology. The Director-General promised in return to help Stutchbury evaluate ores, classify strata and otherwise prosecute his investigations.

Stutchbury began his survey in 1850 but he was upstaged the following year by the decisive gold strike of the prospector Edward Hargraves. The search for valuable minerals now assumed real urgency. While Stutchbury examined the western flanks of the Great Dividing Range, Clarke conducted a special gold survey of the chain's eastern flank and the southern Alps. In the process, he discovered a series of Palaeozoic fossils that, when classified in Britain, confirmed Strzelecki's lithological evidence of an important development of Palaeozoic strata in Australia. This was welcome news to Murchison, who was always eager to extend his domain among the ancient rocks. He and Clarke nevertheless fell out over their rival claims to priority for predicting the gold discovery and their competing theories about the occurrence of the metal. Yet the two men joined forces to defend their science against Hargraves, who had attacked the geologists for failing to precipitate a gold-rush. Public opinion decided this triangular contest. The subordinate cultural status of colony vis-à-vis metropolis is illustrated by the jack-of-all-trades Hargraves triumphing over Clarke in New South Wales as the harbinger of golden prosperity, and the well-connected savant Murchison receiving wide credit in Britain for initiating this profound change in imperial affairs.

The Murchison-Clarke dispute continued to flare periodically, for though Murchison helped Clarke attain recognition and honours in Britain, he used his influence to ensure that he himself received an undue share of credit for what was in actuality an informed guess about the gold. His treatment of Clarke demonstrates how the gentlemen-scientists commanding the institutions of the imperial capital were able to maintain their authority over colonial practitioners. Attempts to enforce subservience rankled emigrant scientists such as Clarke who considered their status equal to that of British colleagues. The dominance of colonial scientists by London savants, however, although partially structural, was not unique to the colonial condition. As David Knight emphasizes, provincial scientists in Britain were treated similarly. Centralized control of the flow of scientific information within the empire was a key means by which the metropolitan élite maintained its ascendency both within the imperial system and on the international front.[25]

The gold-rushes marked a watershed in the history of Australian geology. Clarke's appointment ended in 1853 and Stutchbury's two years later, but their employment encouraged other colonies to follow suit. In 1850, in the course of evaluating rich new-found deposits of copper and lead from Western Australia, De la Beche suggested that a geologist be appointed to that colony. His advice was not acted upon, probably because of the availability there of von Sommer and the Gregory brothers, explorers whose work had been published by the Geological Society of London.[26] In 1851 South Australia requested on the basis of its booming copper mines as well as the gold discoveries that a geologist be sent out by the home Survey. The contract terms stipulated by De la Beche proved too stringent for the colonial Legislative Council, however, and the emigrating civil engineer Benjamin Babbage was recommended to conduct a temporary survey instead.[27]

In Victoria, the gold discoveries of 1852 provided the incentive and the means to establish a geological survey that soon became the model for such institutions throughout the British empire. Alfred Selwyn was the geologist recommended by De la Beche to reconnoitre the colony's mineral resources. He had had seven years' experience with the home Survey, and his work among the gold-bearing Palaeozoic rocks of Wales proved especially relevant in Victoria. Selwyn began surveying the goldfields in 1853 but was soon ordered to search for coal near Melbourne. This proved unsuccessful, and Selwyn found his work handicapped by insufficient staff and a dearth of accurate maps upon which to record his observations.

In 1856 the Victorian Survey was re-established on a more permanent footing. Selwyn was able to increase his staff, accelerate surveying, and produce an impressive series of geological maps. For the most part he trained his assistants on the job, although Christopher Aplin had studied briefly at the School of Mines in London and George Ulrich had graduated from a German mining academy. A third geologist, Richard Daintree, resigned from the Victorian Survey and spent six months at the School of Mines, later being reappointed by Selwyn. The links between the Victorian and British surveys remained close under the Director-Generalship of De la Beche and of Sir Roderick Murchison, who succeeded him in 1855. Specimens and publications were regularly exchanged, and specialists at the Museum of Practical Geology initially helped Selwyn date fossils and evaluate economic minerals.

The fossil work was soon being done in Victoria, however, for in 1854 the palaeontologist Frederick McCoy, who had served two years with the Irish Survey, was appointed to the chair of natural science at the new University of Melbourne. In 1856 McCoy was also appointed Palaeontologist to the Victorian Survey. McCoy used his authority to create a geological and mining collection within the colony's National Museum, the functions of which he envisaged as paralleling those of the British Survey's Museum of Practical Geology.[28] He also supported Selwyn's campaign to ensure that his survey be carried out according to the standards of the home Survey. The addition of a chemical laboratory in 1863, staffed initially by Charles Wood, a graduate of the School of Mines formerly employed on a survey in New Zealand, made the structural resemblance between the colonial and metropolitan establishments more exact.[29]

The methods of the British Survey were thus transmitted to Australia by the direct transfer of personnel. Yet grave problems were to emerge from the attempt to interpret Antipodean geology according to European precedents. Clarke, for example, concluded that the coal deposits of New South Wales were Palaeozoic, as were Britain's Carboniferous seams, while McCoy assigned them a Mesozoic (Jurassic) age. For thirty years the two scientists pursued a bitter debate over an issue that affected public perceptions of the economic future of the Australian colonies, as well as scientific theories concerning climatic and biological history. The calorific quality of the New South Wales coal was well established by this time. But Eurocentric notions regarding the inferiority of Mesozoic coals in comparison with older Carboniferous fuels meant that public opinion as to the apparent capacity of Australia to undertake industrialization could be influenced by scientific pronouncements. Clarke's view that the major deposits of eastern Australian were Palaeozoic, if not specifically Carboniferous, gradually prevailed. The coals in question are Permian, but the eventual proof of the occurrence of Mesozoic seams in the same area demonstrated that Antipodean geology differed from that of Europe in exhibiting no clear break between Palaeozoic and Mesozoic formations.[30]

A second case illustrating the overthrow of European preconceptions involves gold. Murchison believed gold to be peculiarly associated with his own Silurian system, while Clarke maintained that the metal occurred in veins penetrating sedimentary strata of various ages, as well as being diffused in igneous rocks. Yet such was the prestige of Murchison as the foremost geological authority on gold that his views were widely accepted even in Australia. Murchison also believed that gold veins decreased in richness in proportion to their depth. In the early stages of the Victorian gold-rush, field evidence seemed to support Murchison's views: indeed, Selwyn subscribed to them. As deeper sinkings continued to yield paying quantities of gold, however, and capital was invested in quartz crushing companies, the maintenance of a permanent mining industry in the colony began to seem possible. Murchison, as Director-General, railed against such undertakings. He was supported by McCoy, who discouraged investment in deep-level quartz mines and in railways to service these centres. Victoria's miners were outraged and Governor Sir Henry Barkly feared that an economic slump might result from such pessimistic publicity. In consequence Selwyn, who had already rejected Murchison's theory, was prompted by Barkly to rebut the position of

McCoy and Murchison. The investors were thereby reassured, and Murchison was forced to acknowledge that Australian gold failed to conform to theories conceived in London.[31]

Selwyn was defending his science, not merely protecting the profits of Victoria's mine owners. It is tempting to view his rebellion against Murchison as a declaration of scientific independence from the authority of the metropolis. But Murchison was compelled to alter his views precisely because of Selwyn's impeccable metropolitan credentials. Both men were in fact British imperial geologists, rather than representatives of rival colonial and metropolitan interests. Indeed, during this era metropolitan containment of scientific developments at the periphery paralleled political envelopment through Whitehall's granting of responsible government to the colonies.[32]

As Director-General, Murchison continued to support the Victorian Survey and to use its example to press for the installation of geological surveys in other colonies. He also arranged to supply Selwyn with promising graduates of the Royal School of Mines: Henry Brown was thus recruited in 1865. Bureaucratic controversy, however, led to the disbandment of the Victorian Survey in 1869. Selwyn, in a move further illustrating the imperial nature of his work, then became Director of the Geological Survey of Canada. The staff he had trained was also dispersed to other colonies. Ulrich became professor of mining and mineralogy at the Otago School of Mines in New Zealand. Brown worked on surveys in Western Australia, New Zealand, Canada, and New South Wales before concluding his career in South Australia. Aplin and Daintree were temporarily appointed as government geologists in Queensland in the late 1860s, rendering that colony the first Australian territory to be examined by geologists whose primary professional training had been acquired in Australia.[33]

The mainland gold-rushes also excited expectations in Tasmania, and in 1856 Murchison was requested to recommend a geologist to survey that island's mineral resources. Charles Gould, a son of the ornithologist John Gould and another graduate of the School of Mines, was chosen for the post. Gould served in Tasmania between 1859 and 1869, but he made no major economic discoveries. He accomplished a good deal of original exploration, however, sketched out the stratigraphy of most of the island, and contributed evidence on the age of Australian coal.[34]

McCoy's appointment demonstrates that British recommendations also dominated the foundation of academic geology in Australia. At the

University of Sydney, the first reader in geology was Alexander Thomson, chosen by the ubiquitous Murchison. When Thomson died in 1872, he was replaced by Archibald Liversidge, a graduate of the Royal School of Mines who was to play a key role in the development of Australian science. Liversidge established the University of Sydney's Faculty of Science in 1882 and, with the backing of his colleague T.W. Edgeworth David and the New South Wales Department of Mines, helped found the School of Mines a decade later. David, professor of geology at Sydney from 1891, was an Oxford graduate who had eight years' experience with the Geological Survey of New South Wales. David's contributions to Australian stratigraphy and glaciology were enormous, and several of his students went on to distinguished careers throughout the empire. Other professors of geology – J.W. Gregory at the University of Melbourne and Ralph Tate in the chair of natural science at Adelaide – were recruited directly from Britain. Ernest Skeats, who replaced Gregory in 1905, had graduated from the Royal School of Mines and the University of London. The London committees that recommended Gregory and Skeats were in both cases entirely British in composition. David's appointment departed from this pattern, for he was selected by a local committee that rejected the recommendation of the metropolitan advisers. Until well beyond the turn of the century, however, the imperial connection remained strong in Australian academic geology.[35]

The history of survey geology reveals a similar pattern. Work on Australian fossils accomplished by members of the Geological Survey of India established the stratigraphic correlation between Australia and India that made possible the correct dating of the Australian coals and eventually led to the Gondwana concept of an ancient unified southern continent.[36] By 1890 all of the Australian colonies were endowed with permanent geological surveys. Victoria's was reformed in 1870, and by 1900 a broad outline of the colony's geology and mineral deposits had been obtained. The Geological Survey of New South Wales was re-established in 1875 by Charles Wilkinson, another of Selwyn's men. Robert Etheridge, Junior, the son of the distinguished British palaeontologist of the same name, had worked as an assistant geologist on Selwyn's survey between 1866 and 1868. He had then served on the Scottish Survey and at the British Museum before receiving a joint appointment in 1887 as Palaeontologist to the Geological Survey of New South Wales and the Australian Museum. Etheridge's work, together with that of Tate, helped to free Australian palaeontology from its European theoretical bondage. The graptolite studies of Thomas Sergeant Hall, a pupil of McCoy who lectured at the University of Melbourne between 1894 and 1915, brought stratigraphic order to the lower Palaeozoic rocks of Victoria. Hall's work provided a basis for the dating of these early strata throughout the southern hemisphere, and he became the first Australian palaeontologist of truly international distinction.[37]

In 1876 Queensland appointed as its government geologist Robert Logan Jack, another veteran of the Scottish Survey. In 1883 he was joined by William Rands, a graduate of the Royal School of Mines who eventually succeeded him. Jack spent 22 years surveying Queensland, and he explored much of the Cape York Peninsula as well as conducting a reconnaissance in British New Guinea.[38] Australia's expansionist ambitions during the *fin-de-siècle* era of imperial rivalry were also reflected in geological explorations in Antarctica and various Pacific islands. Edgeworth David's 1897 expedition to the atoll of Funafuti to test Darwin's theory of coral reef formation – an enterprise conducted in competition with a British expedition of the previous year – provides another example of the 'sub-imperialism' of Australian science in this period.[39] Jack himself went even further afield in private service connected with British imperial interests. Having been appointed Government Geologist of Western Australia in 1898, he resigned the following year in order to explore Szechwan province for a British firm that had secured a mining concession from the Chinese government. Britain, anxious not to be shut out in the anticipated dismemberment of China, backed this venture, but the Boxer Rebellion prevented development of the ore deposits that Jack discovered.[40]

The Tasmanian Geological Survey was refounded in 1882 after tin, copper, and gold discoveries ignited fresh interest in mining. In the same year South Australia founded a permanent survey under Henry Brown, formerly of Selwyn's Victorian Survey. Until his retirement in 1912, Brown explored the geology of Central Australia and provided the first systematic knowledge of the Northern Territory. Western Australia had employed Brown between 1870 and 1872, but another government geologist was not appointed until 1882, when E.T. Hardman of the Irish Survey took up the post. Hardman accompanied two exploring expeditions to the Kimberley region, and his assessment of the gold potential of this remote area sparked Western Australia's first gold-rush in 1886. Hardman returned to Ireland in 1885 when the colonial government could no longer afford his services. The Reverend Charles Nicolay, an English clergyman and geographer who founded Perth's first public museum, partially filled the resulting vacancy by advising the colonial government on geological issues. In 1887 a permanent geological survey was at last established in Western Australia.[41]

By this time, survey geologists were being appointed from within Australia and were as frequently Australian-born as British. Nevertheless, metropolitan influence remained strong. The descriptive work accomplished by the geological surveyors, it should be emphasized, was of both economic and theoretical importance. Throughout the nineteenth century, Australian geology survived largely as a creature of government patronage. Yet the mid-century gold-rushes and the subsequent development of major deposits of other metals, such as the silver-lead lode of Broken Hill in New South Wales and the Mount Lyell copper mines of Tasmania, were initiated by prospectors and mining engineers. Few consulting geologists were employed in Australia before the Second World War, and specialized examinations for private firms were usually carried out by visiting experts from Britain or the United States.[42]

The history of geology is a study of ideas, individuals, and institutions rather than of rocks alone. The cultural context in which Australian geology took root and flourished was overwhelmingly British. Vallance has put the case that the discipline as practised in Australia remained dominated by European concepts until the mid-1870s.[43] The evidence suggests, however, that in terms of personnel, theory formulation, publication and primary societal affiliation, Australian geology remained largely an extension of British geology until well past the turn of the century. Australian geologists likewise continued to seek their postgraduate qualifications from British universities as late as the 1960s. Roy MacLeod maintains that Britain's leadership of science in the Empire/Commonwealth continued until long after the Second World War.[44] But whatever date may be assigned for a gradual divergence that, like continental drift, is in the case of geology still occurring, the dependent status of Australian science, like many other aspects of Australian cultural and economic life, lasted far beyond the point of ostensible political independence marked by Federation in 1901.

Eurocentrism certainly distorted the progress of Australian geology, as the cases of coal and gold illustrate. Throughout most of the nineteenth century, educational institutions capable of providing specialized geological training were non-existent in Australia. When university geology departments and schools of mines were established in the third quarter of the century, they were mainly staffed by Britons. Under these circumstances it was inevitable that geologists in Australia would initially rely on European theories, models and techniques in classifying the structure of new environments because they lacked a local data base. Modifications eventually followed as field evidence failed to agree with the imported interpretive framework. As the inhibiting influences were overcome, an indigenous geology appropriate to Australia developed. This process of import substitution also gradually transformed the field of mining engineering in Australia.

It must nevertheless be emphasized that British inputs were crucial to this process. In the early days of colonization, the research agenda of most geologists working in Australia was directly determined in London. Even when colonial governments began paying the salaries of survey and academic geologists in Australia, such posts emulated British models and were frequently held by products of those parent establishments who came out to Australia as cultural emissaries bearing a full complement of metropolitan assumptions and theoretical baggage. Since the knowledge these geologists were expected to generate was intended to benefit not only the colonies but the mother country's economy and prestige as an imperial power, it can be argued that nineteenth-century Australian geology constituted nothing more than British imperial geology as performed at one particular locus. Paradoxically, however, it was agents of the metropolis who presided over the transformation of colonial geology into a recognizably Australian geology; just as the larger cultural tapestry, of which this science formed one strand, was gradually transformed.

Accruing evidence forced such a reorientation in the interpretation of phenomena. But in its organizational and orientational aspects the need for the Antipodean version of the discipline to diverge from its metropolitan parent was less compelling. Given the pre-eminence of British stratigraphic geology during most of the nineteenth century – a pre-eminence achieved and maintained partially as a result of privileged access to imperial research opportunities – there was good reason for Australian geology to preserve close ties with its British forebear. In aspects of the science such as petrology, in which the British achievement was less than outstanding, the incentives for colonial emulation were not so compelling. Yet it is clear that the overall subordination of Australian geology to the imperial capital was actively maintained at both the institutional and individual levels. As the essays of Rod Home and

Elizabeth Newland testify, the maintenance of dependence between colonial and metropolitan science required both the exercise of patronage and control on the part of British savants and active collaboration by emigrant scientists. Psychological and social factors, as well as institutional and geographical ones, require consideration in analysing such relationships.

Imperial science was a multilateral arrangement benefiting all the contracting parties. It was an unequal exchange, however, in that the central player received the lion's share of the available rewards. With Britain as the prime mover and largest investor in the imperial game, it was not unreasonable that the profits should be weighted in her favour during the initial stages of play. As a daughter society, however, Australia received immense reciprocal benefits roughly commensurate with her changing level of development. The balance of relative inputs and outputs has shifted, over time, with changes in the international political and economic equations, and the game of empire in its original guise has all but vanished from the board. Australian geology is now independent, mature, and fully functional in its own right. But the British empire provided the setting in which the discipline developed in the Antipodes and the means by which it was exported from Britain to colonial loci throughout the world. The Australian experience has thus paralleled that of other colonies within the British empire, which served as a nursery of scientific traditions as well as of nations.

In contrast to the United States, which enjoyed an internal scientific frontier in the nineteenth century, Britain's lay overseas on the far-flung borders of the empire. Australia and New Zealand were the most distant of these frontiers and, in terms of geology, the most challenging in their difference from European conditions. British geologists enjoyed peculiar advantages in Australasia because the region constituted an exclusive British preserve where research was facilitated by the familiar colonial culture and infrastructure. For Victorian British geologists, Australia thus represented a quarry for data and a market for jobs, much as the country has remained a quarry for foreign consumers of natural resources and a market for the finished goods manufactured from these primary products. British scientists also viewed the empire as a gigantic laboratory. There were marked similarities between colonial and provincial science. Yet being at the periphery could prove advantageous in that frontiers could function as forcing houses for innovation as readily as they might constitute stagnant backwaters fostering second-rate, derivative endeavour.

Many of the geologists who came to Australia from Britain in the nineteenth century would not have made the shift had not financial, career, or health considerations compelled them to do so. Those who did come tended to be marginal men driven to the periphery by the exigencies of professional science in the metropolis. We have only to recall that Jukes as well as Bristow turned down the post in New South Wales before Stutchbury accepted it, or to contemplate that it required four years to find a geologist willing to go to Tasmania, to realise what even a temporary transfer to Australia meant to early- and mid-Victorian geologists. Psychologically, acceptance of a post in the southern colonies was akin to a sentence of transportation. The constant complaints to colleagues at home about field conditions, unrealistic official expectations, philistine colonial attitudes and intellectual isolation evince a sense of exile. Only rarely, as with Jukes or Clarke, did a British geologist write of the exhilaration of working in a strange and appealing environment. But even Clarke would have returned to Britain if given the chance, while marriage and promotion to the Directorship of the Irish Survey were sufficient to keep Jukes at home.

From the West Indies to Cape Colony to India, the attitude was much the same. Despite the attraction of virgin fields for research, geologists went reluctantly, usually lured by lucrative salaries, and longed to return to Britain for as long as they sojourned in the colonies. These circumstances could breed the cynicism of the 'carpetbagger scientist' who sought to exploit the colonial setting for its high wages, professional opportunities and untapped data before returning to Britain to deploy this looted cultural capital in the interests of personal advancement. Colonial conditions could also produce among emigrant scientists the garrison mentality endemic to outposts of empire. This malady could be felt as acutely by Australian academics like McCoy as by officers of the Indian Army or factors of the Hudson's Bay Company.[45]

The southern colonies were a mirror for scientists as well as emigrants, reflecting and distorting the British scene. Some, like Selwyn or Babbage, were able to follow Lewis Carroll's heroine and pass through the looking glass into the new world of Australia. Others, like Gould, saw only an ill-formed and inferior copy of the Britain they had left. Collectively, however, the British geologists who worked in Australia made substantial contributions to knowledge of the structure of the country and the nature of its mineral resources. Culturally and economically, they thus facilitated

Australia's development and its integration into the European world system dominated in the nineteenth century by Britain. They founded institutions and patterns of research that continued to operate within the British tradition, and they ascertained the location and nature of mineral deposits the exploitation of which continues to underwrite a large percentage of the Australian economy. Above all, these geologists helped Europeans to come to terms with what seemed a very alien environment. By classifying Australia's rocks and fossils according to the time-frame provided by international stratigraphy, they helped make sense in European terms of its landscape and life forms. As they did in other territories being opened up throughout the empire, British geologists helped the colonists control Australia by giving their new land physical and historical dimensions. In intellectual as well as practical terms, British geologists working in Australia were active agents of imperialism.

Department of History
La Trobe University
Australia

NOTES

Abbreviation used: *ADB* = *Australian Dictionary of Biography*

1 C.A. Browne, ed., *Letters and Extracts From the Occasional Writings of J. Beete Jukes* (London, 1871), pp. 264–5.

2 Charles Lyell, *Principles of Geology*, 3 vols. (London, 1830–33), I, 56–57.

3 R.I. Murchison, 'Address to the R.G.S.', *Journal of the Royal Geographical Society*, 27 (1857), xciv–cxcviii; p. cxxxvii.

4 Martin J.S. Rudwick, 'The Foundation of the Geological Society of London: Its Scheme for Co-operative Research and Its Struggle for Independence', *British Journal for the History of Science*, 1 (1963), 325–355.

5 See J.B. Morrell, 'Economic and Ornamental Geology: the Geological and Polytechnic Society of the West Riding of Yorkshire, 1837–53', in I. Inkster and J.B. Morrell, eds., *Metropolis and Province: Science in British Culture, 1780–1850* (London, 1983), pp. 231–256; Robert A. Stafford, 'Geological Surveys, Mineral Discoveries, and British Expansion, 1835–71', *Journal of Imperial and Commonwealth History*, 12 (1984), 5–32; idem, *Scientist of Empire: Sir Roderick Murchison, Scientific Exploration, and Victorian Imperialism* (Cambridge, 1989); cf. Roy Porter, 'Gentlemen and Geology: the Emergence of a Scientific Career, 1660–1920', *Historical Journal*, 21 (1978), 809–836. For Continental developments, see Rachel Laudan, *From Mineralogy to Geology: the Foundations of a Science, 1650–1830* (Chicago, 1987).

6 David Mackay, *In the Wake of Cook: Exploration, Science, and Empire, 1780–1801*

(London, 1985).

[7] Paul J. Weindling, 'Geological Controversy and Its Historiography: the Prehistory of the Geological Society of London', in L.J. Jordanova and Roy Porter, eds., *Images of the Earth: Essays in the History of the Environmental Sciences* (Chalfont St. Giles, 1979), pp. 248–271.

[8] Morris Berman, *Social Change and Scientific Organization: The Royal Institution, 1799–1844* (London, 1977).

[9] P.J. Cain and A.G. Hopkins, 'Gentlemanly Capitalism and British Expansion Overseas. I: The Old Colonial System, 1688–1850', *Economic History Review*, 2nd ser., 39 (1986), 501–525; 'II: New Imperialism, 1850–1945', 40 (1987), 1–26.

[10] James A. Secord, 'King of Siluria: Roderick Murchison and the Imperial Theme in Nineteenth-Century British Geology', *Victorian Studies*, 25 (1982), 413–442; Stafford, op. cit. (n. 5, 'Geological Surveys').

[11] T.G. Vallance, 'The Start of Government Science in Australia: A.W.H. Humphrey, His Majesty's Mineralogist in New South Wales', *Proceedings of the Linnean Society of N.S.W.*, 105 (1981), 107–146.

[12] J.S. Flett, *The First One Hundred Years of the Geological Survey of Great Britain* (London, 1937); E.B. Bailey, *The Geological Survey of Great Britain* (London, 1952); Paul J. McCartney, *Henry De la Beche: Observations on an Observer* (Cardiff, 1977); Margaret Reeks, *Register of the Associates and Old Students of the Royal School of Mines, and History of the Royal School of Mines* (London, 1920). For Ireland, see G.L. Herries Davies, *Sheets of Many Colours: the Mapping of Ireland's Rocks, 1750–1890 (Dublin, 1983).*

[13] James A. Secord, 'The Geological Survey of Great Britain as a Research School, 1839–1855', *History of Science*, 24 (1986), 223–75.

[14] David F. Branagan, 'Phillip Parker King: Colonial Anchor Man', in A. Wheeler and J. Price, eds., *From Linnaeus to Darwin: Commentaries on the History of Biology and Geology* (London, 1985), pp. 179–93.

[15] T.G. Vallance, 'Presidential Address: Origins of Australian Geology', *Proceedings of the Linnean Society of N.S.W.*, 100 (1975), 13–43.

[16] Joseph B. Jukes, *Sketch of the Physical Structure of Australia* (London, 1850).

[17] Kathleen Dugan, 'Darwin and Diprotodon: the Wellington Caves Fossils and the Law of Succession', *Proceedings of the Linnean Society of N.S.W.*, 104 (1979), 265–72.

[18] Michael Hoare, ' "All Things Are Queer and Opposite" – Scientific Societies in Tasmania in the 1840's', *Isis*, 60 (1969), 198–209.

[19] Robert A. Stafford, 'The Long Arm of London: Sir Roderick Murchison and Imperial Science in Australia', in R.W. Home, ed., *Australian Science in the Making* (Sydney, 1988), 69–101; David F. Branagan, 'Strzelecki's Geological Map of Southeastern Australia: an Eclectic Synthesis', *Historical Records of Australian Science*, 6 (3) (1986), 375–92.

[20] Vallance, op. cit. (n. 15); for Menge, see also Bernard O'Neil, *In Search of Mineral Wealth: the South Australian Geological Survey and Department of Mines to 1964* (Adelaide, 1982), pp. 8–19.

[21] Vladislav Kruta et al., *Dr John Lhotsky, the Turbulent Australian Writer, Naturalist, and Explorer* (Melbourne, 1977); Gerda Bell, *Ernst Dieffenbach: Rebel*

and Humanist (Palmerston North, 1976).
[22] T.G. Vallance, 'Early German Connections with Natural History, Geology and Mining in New South Wales and Queensland', in J.H. Voigt, ed., *New Beginnings: Germans in New South Wales and Queensland* (Stuttgart, 1983), 269–78.
[23] James Jervis, 'Rev. W.B. Clarke, "The Father of Australian Geology"', *Royal Australian Historical Society, Journal and Proceedings*, 30 (1944), 345–458; Elena Grainger, *The Remarkable Reverend Clarke: The Life and Times of the Father of Australian Geology* (Melbourne, 1982); Stafford, op. cit. (n. 19).
[24] Ann Mozley, 'The Foundation of the Geological Survey of New South Wales', *Journal and Proceedings of the Royal Society of N.S.W.*, 98 (1965), 91–100; *ADB*, vol. 6, pp. 216–217.
[25] For Murchison versus Clarke, see Stafford, op. cit. (n. 19); for Murchison's role as a mediating authority in geological affairs throughout the empire, see Stafford, op. cit. (n. 5, *Scientist of Empire*).
[26] N.W. Archbold, 'Western Australian Geology: an Historical Review to the Year 1870'. *Journal of the Royal Society of Western Australia*, 63 (1981), 119–128.
[27] *ADB* , vol. 3, pp. 65–66; O'Neil, op. cit. (n. 20), pp. 25–31.
[28] Sally G. Kohlstedt, 'Australian Museums of Natural History: Public Priorities and Scientific Initiatives in the Nineteenth Century', *Historical Records of Australian Science*, 5 (4) (1983), 1–29.
[29] Thomas A. Darragh, 'The Geological Survey of Victoria under Alfred Selwyn, 1852–1868', *Historical Records of Australian Science*, 7 (1) (1987), 1–25; idem, 'Charles Sturtevant Wood, Analyst, his Work and his Correspondence with James Hector', ibid., 7 (3) (1988), 231–71; Stafford, op. cit. (n. 19).
[30] T.G. Vallance, 'The Fuss About Coal: Troubled Relations Between Palaeobotany and Geology', in D.J. and S.G.M. Carr, eds., *Plants and Man in Australia* (London, 1981), 136–76.
[31] Stafford, op. cit. (n. 19).
[32] Roy MacLeod, 'On Visiting the "Moving Metropolis": Reflections on the Architecture of Imperial Science', *Historical Records of Australian Science*, 5 (3) (1982), 1–16.
[33] R.K. Johns, ed., *History and Role of Government Geological Surveys in Australia* (Adelaide, 1976).
[34] *ADB*, vol. 4, pp. 277–278.
[35] D.F. Branagan and K.A. Townley, 'The Geological Sciences in Australia – a Brief Historical Review', *Earth Science Reviews*, 12 (1976), 323–346; for Thomson's appointment, see also Stafford, op. cit. (n. 19); for Gregory, see David Branagan and Elaine Lim, 'J.W. Gregory, Traveller in the Dead Heart', *Historical Records of Australian Science*, 6 (1) (1984), 71–84.
[36] Vallance, op. cit. (n. 30).
[37] T.G. Vallance, 'Pioneers and Leaders: a Record of Australian Palaeontology in the Nineteenth Century', *Alcheringa*, 2 (1978), 243–50; Libby Robin, 'Thomas Sergeant Hall (1858–1915): Scholar and Enthusiast', *Historical Records of Australian Science*, 6 (4) (1987), 485–92.
[38] Johns, op. cit. (n. 33).
[39] Roy MacLeod, 'Imperial Reflections in the Southern Seas: the Funafuti Expedi-

tions, 1896–1904', in Roy MacLeod and Philip F. Rehbock, eds., *Nature in its Greatest Extent: Western Science in the Pacific* (Honolulu, 1988), pp. 159–191.
[40] W.F. Collins, *Mineral Enterprise in China* (London, 1918).
[41] Johns, op. cit. (n. 33), passim.; for Hardman, see Geoffrey Blainey, *The Rush That Never Ended: A History of Australian Mining*, 2nd ed. (Melbourne, 1969), pp. 161–68; for Nicolay, *ADB*, vol 5, pp. 340–41.
[42] Branagan and Townley, op. cit. (n. 35), 336–8.
[43] Vallance, op. cit. (n. 15).
[44] MacLeod, op. cit. (n. 32).
[45] Barry Smith, 'Stalwarts of the Garrison: some Irish Academics in Australia', *Australian Cultural History*, 6 (1987), 74–93.

MICHAEL A. OSBORNE

A COLLABORATIVE DIMENSION OF THE EUROPEAN EMPIRES:

*Australian and French Acclimatization Societies and Intercolonial Scientific Co-operation**

On 10 May 1883, Ferdinand Mueller, the most famous Australian Scientist of his era, informed the Acting Consul General for France that he was reluctant to send seeds of native Australian plants to an institute of agronomy in the French colony of Algeria. For 'the last 25 years and even more', wrote Mueller, 'I have made so many sendings of seeds to Algeria, that all our ordinary trees and shrubs have become common there'. Unless extremely rare varieties of Australian plants were needed, Mueller continued, it would be easiest to buy the seeds from local merchants in Algiers.[1] Mueller's letter raises an important and little-explored feature in the history of science and nineteenth-century European colonialism – scientists in the colonies collaborating *across* the political boundaries imposed by their confederations to assist in the development of colonies under the control of other European nations.

Collaborative internationalism does not exclude nationalism, and to be sure, it is also important to examine the changing relationship between the styles of science practised in the colonies and the shifting sands of metropolitan politics *within* the same Empire, be it British, French, or German. However, Roy MacLeod, who has shown just how fruitful such an approach can be, has recently pointed out how the interpretation of the history of Australian science might transcend Australian history.[2] MacLeod is not alone in raising the international dimension of colonial science, and during the last decade and a half, others have subjected the history of colonial science to comparative analysis. While internationalist perspectives are explicit in the contributions of Joseph M. Powell on the historical geography of settlement, in the works of Susan Sheets-Pyenson and Sally Gregory Kohlstedt on museums, in Lewis Pyenson's account of the exact sciences in the German Empire, and in Alfred Crosby's study of the history of ecosystems, these authors have given little play to the theme of collaboration across empires by colonial scientists.[3] Nonetheless, their work forms the point of departure for a history of science and colonization that cuts across empires.

R.W. Home and S.G. Kohlstedt (eds.), International Science and National Scientific Identity, 97–119.

During the middle third of the nineteenth century, scientific societies arose in both colony and metropolis. Many of these groups undertook to transplant exotic flora and fauna, and a number of them – the acclimatization societies – made these transplantations, ostensibly in accord with scientific principles, their *raison d'être*. Acclimatization societies sprang up in all the great European empires. The present study examines the history of these colonial groups and their relationship with their metropolitan counterparts and links to each other in the French and British empires. To make a global topic manageable, the discussion concentrates on the efforts of voluntary scientific societies concerned with the acclimatization of animals. That being said, however, these organizations, like many others in the colonies, were polyvalent entities of limited lifespan, and in practice they assumed responsibility for all topics concerning acclimatization – animal, plant, and human.

A further delimiting of topic sets aside India. Both the French and British were active in south-east Asia, and the wisdom of excluding India from any discussion of nineteenth-century British colonialism ought to be questioned. However, much of the plant acclimatization done there was directed by Kew Gardens or the East India Company rather than by voluntary scientific societies.[4] Moreover, comparing the history of a settler colony like Australia with that of India is a bit like comparing apples and oranges. Neither the French nor the British developed their Indian holdings as settler colonies. Pondicherry remained a sort of trading station and India became an arena for plantation agriculture. Hence, in the interest of symmetry, and because colonial acclimatization society activity was most intense in Australasia and French North Africa, we are left with the two glamour settlement colonies of the two great European empires of the nineteenth century, Algeria and Australia.

Without doing violence to that which is autochthonous in the various colonies, or for that matter to the unique features of French and British science, a comparison of science and its institutions in nineteenth-century Australia and Algeria suggests that the economic and social conditions found in the colonial cities allowed certain kinds of science to take root and flower during certain eras. There were in fact many real similarities between the socio-economic environments of Melbourne and Algiers. Both Australia and Algeria became settlement colonies and hence they occupied similar niches within their respective empires. At mid-century the French had occupied Algeria for two decades. They looked towards the antipodes with good reason, for by all accounts, they had not made much of their own colony. French agricultural scientists admired, studied in detail, and sought to emulate the success of the British in Australia. By the late 1860s, largely by planting trees such as the Australian eucalyptus and draining swamps, the French had so improved the sanitary state of Algeria that British study commissions began to extol it as a model for hygienic science in India.[5] Of crucial importance for the history of agronomy and economic zoology in French Algeria – of which acclimatization societies were an important part – French scientists perceived the climate of Algeria as being very similar to Australia.

Historical understanding of scientific societies and their role in European culture has matured of late, and we now have case studies of several societies and a few attempts at synthesis.[6] As the work of Michael Hoare and Linden Gillbank has shown, Europeans in the antipodes actively participated in the corporate life of science.[7] The acclimatization society movement, which crested in both the French and British empires in the 1860s and 1870s, is actually one significant stream within a larger cultural current. The acclimatization societies aided in the diffusion of science, engaged in what they saw as patriotic and utilitarian projects, and, of course, their members collected the natural historical productions of their environment for local and metropolitan museums. Certainly these groups served as forums for the socially ambitious, but their main function was to import and establish foreign organisms.

Except for occasional asides, the following remarks apply to a collage of activities and individuals associated with five groups dedicated to animal acclimatization: the *Société zoologique d'acclimatation* founded in Paris in 1854, and that society's branch in Algiers, the *Comité d'acclimatation de l'Algérie*, founded in 1859; the Society for the Acclimatisation of Animals, Birds, Fishes, Insects, and Vegetables Within the United Kingdom, founded in London in 1860; and two Australian societies founded in 1861, Melbourne's Acclimatisation Society of Victoria, and Sydney's Acclimatisation Society of New South Wales. For convenience, these groups will here be referred to by the municipality in which they were founded. The Algiers group, which had about one hundred people present at its first meeting, was properly a branch of the Paris society. The Melbourne society, possibly the largest and wealthiest acclimatization society in the British Empire, and the more modest group in Sydney, were independent of the London society which they outlived.

Sometimes the Algiers and Australian groups collaborated via Paris, often through the French Consul in Sydney.[8] Although the Paris society functioned as the global clearing house for animal acclimatization, it fell far short of directing the wide range of acclimatization activity. Key members of the various groups, rather than the organizations *per se*, often acted together on personal initiatives for the cause of acclimatization. This informal collaboration was especially common among those members who directed scientific institutions. Isidore Geoffroy Saint-Hilaire, the first president of the Paris society and director of the menagerie at the Paris Museum of Natural History, built the group's power base within the scientific community by packing its ranks with museum curators and zoo directors. Affiliation with an acclimatization society rationalized the activities of some, but individuals such as Ferdinand Mueller, director of the Melbourne Botanic Gardens, Auguste Hardy, director of the Jardin d'essai at Algiers, and Charles-Victor Naudin, director of the agronomic institute at Antibes, France, had a good deal of autonomy. These acclimatization society members and others often exchanged materials, performed experiments, and then published the results under the banner of the Paris society, which gave its blessing or awarded prizes after the fact or while experiments were *in medias res*.

International collaboration was vital to the acclimatization societies because their work necessitated a global horizon, one that reached beyond the confederations established by the mother country. The acclimatization societies of Australasia and Algiers were not just interested in exchanging organisms with one another. These groups tried to stimulate industries founded on organisms new to both North Africa and Australia, and often they were the same organisms. For example, the basic ingredients of llama culture and sericulture, llamas and silkworms – two of the industries promoted by acclimatization societies in Algeria and Australia – were to be found in Latin America and Asia. The Paris group frequently obtained easily transportable materials like silkworms and bamboo at little or no cost. Its numerous members in the French diplomatic corps carried lists of desiderata to their posts of duty where they collected materials requested by the group. Charles de Montigny, the first French diplomat stationed in Shanghai, recruited many Asian residents to the Paris society and sent back numerous cases of silkworm eggs, live birds, and even a herd of yaks. Sir Rutherford Alcock, the ranking British diplomat in Asia and Montigny's birding companion in Shanghai before he moved to Japan, assisted the Paris society and became an officer.[9] The French viewed Japan as a potential source of new industries for Algeria, and Napoleon III commissioned an officer in the Paris society, F.E. Guerin-Méneville, to establish a silk industry in Algeria with Japanese silkworms.[10] The organisms acquired in Asia usually rode military ships back to France, where they became available for experiment or exchange or found their way into the group's zoo and showcase for economic zoology, the Jardin zoologique d'acclimatation in Paris.

Participation in the 1855 Exposition Universelle publicized the Paris group's existence and brought many foreign collaborators within its orbit. One of the more prominent Australians in Paris in 1855, the exposition commissioner from New South Wales, Sir William Macarthur, joined the Paris society that year and was later elected first vice-president of the Acclimatisation Society of New South Wales. Macarthur was an ideal collaborator for the Paris society. A horse breeder of some note, he was also involved in establishing viniculture in Australia and served as a trustee of the Australian Museum. An agricultural improver, this president of the Agricultural Society of New South Wales was also in frequent contact with the Queensland Acclimatisation Society. He served for several years as a delegate for the Paris group, which charged him to distribute its news to other Australian members, local scientific and agricultural societies, and sectors of the Australian public interested in science.[11]

Procurement of exotic animals and plants was not always free from expenses, and the wealthy Paris and Melbourne acclimatization societies contracted for scarce materials. In the 1850s, the Paris group spent several thousand francs to remunerate Montigny for shipping his yak herd from Siam. Eugène Roehn, who supplied animals to acclimatization societies in France, Australia and Algeria, was one of the more successful animal merchants. During the heady days of the Second Republic, Roehn had tried to form a corporation to import llamas from Latin America. Endorsed by Geoffroy Saint-Hilaire and other scientists, Roehn had promoted the importation of llamas as a kind of utopian solution to French social and agricultural problems. The scheme called for the formation of a limited partnership of investors from Algeria and the south of France, which never materialized. However, Roehn eventually sold some llamas that ended up in the antipodes, and in 1859 he obtained about four dozen llamas and alpacas for the Paris group, an effort that won him 2,000 francs in the society's 1860 prize competition.[12]

METROPOLITAN AND COLONIAL ACCLIMATIZATION SOCIETIES

Although a few scientists, landholders and diplomats in Britain, France, Australia and Algeria were engaged in acclimatization projects at mid-century, the era of the acclimatization society movement effectively opened in 1854 with the foundation of the *Société zoologique d'acclimatation* in Paris, the first scientific society to focus solely on acclimatization. Private collectors in Britain such as the Earl of Derby had made great strides to acclimatize exotic quadrupeds. But as Dr George Bennett, a secretary of the Acclimatisation Society of New South Wales, pointed out, the Earl had died just on the brink of success, and his several years of effort and his menagerie at Knowsley died with him. The history of Knowsley, argued Bennett, was 'a sufficient lesson of the uncertainty of private collections, and the necessity of founding societies based on a general and public interest'.[13]

The *Société zoologique d'acclimatation* grew out of the Paris Museum of Natural History, and its top offices often went to Museum professors such as Isidore Geoffroy Saint-Hilaire, Quatrefages de Bréau, and the two Becquerels, Antoine César and his son Alexandre Edmond. The group's links to the program of transformist zoology elaborated by Geoffroy Saint-Hilaire and Quatrefages de Bréau were explicit, and its ties to the colonies and the international community remained steadfast throughout the Second Empire. After the first president, Isidore Geoffroy Saint-Hilaire, died in 1861, his place was taken by Drouyn de Lhuys, Napoleon III's Minister of Foreign Affairs who proceeded to pack the society with subordinates from his own ministry as well as diplomats of foreign nations.[14] The Melbourne and Sydney acclimatization societies also acted quickly to establish ties with foreign diplomats, perhaps the most vital of all links in the networks established to collect, transport and distribute animal resources.[15]

While Franco-British and Franco-Australian collaboration is a feature of acclimatization society activity, the acclimatization movement grew most successfully in areas of French scientific authority. As many as thirty-nine French metropolitan scientific societies and twenty foreign associations claimed to be under the umbrella of the Paris society, yet its family of formally affiliated groups never extended to the British empire. Of those nations with acclimatization societies formally associated with the Paris group, Switzerland had five while Italy, Russia, Spain and Prussia had two each. In the French colonies, branches of the Paris group and local acclimatization committees joined together in a loose federative model that facilitated the exchange of organisms and information. By 1862, groups functioned in Algeria, French Guyana, Réunion, Martinique and Guadeloupe. In many cases the core participants in these colonial groups had some knowledge of the transformist zoological theories that underlay the French approach to acclimatization. They were often former students of Isidore Geoffroy Saint-Hilaire's lectures at the Paris Museum of Natural History. Cosmopolitan by the very nature of their professions, many left Paris and became directors of colonial gardens, civil servants at the prefecture, physicians, missionaries and diplomats.

The Paris society took a vital interest in the agricultural development and human settlement of Algeria, and one of its first acts was to create a permanent commission to deal with the problems of animal, plant and human acclimatization in the colony. When the group had funds in the 1850s and the 1860s, it sponsored missions to the colony. A typical mission was undertaken by the president of the permanent commission on Algeria, Richard du Cantal. Cantal, who also kept yaks on his cattle farm, held degrees in human and veterinary medicine. His travels included four years' military service in Algeria, where between battles he found time to study the colony's agriculture and domestic animals. In 1857 the Paris society sent him to Algeria on a mission to investigate 'the influences exerted on [animal] races as on individuals by domestication, climate, labour, hygienic conditions, [and] especially by nutrition, [that] powerful agent of the modification of species'.[16] Cantal returned to Paris with a plan for Algerian development that rested on the establishment of pure races of exotic quadrupeds, including llamas and Angora goats, both of which were tried in the colony.[17] The same species of organisms and the strategy to establish pure races found application in Australia, where in 1868 the Acclimatisation Society of Victoria had at least 200 Angora goats.[18] Similar experiments in the two colonies attempted to establish silkworms, imported fish, and ostrich farms.[19]

Of the more than 2,600 people who joined the Paris society during its first seven years of existence, 1854 to 1860, about five per cent listed Algerian addresses. Thus, it was not surprising that this group, clustered in and around Algiers, banded together to form a committee of acclimatization in 1859. Charles Géry, the Prefect of Algiers who founded the committee, functioned as its delegate to the Paris society's council and allowed the group to meet at the prefecture. During the 1860s, the group monitored experiments for the Paris society and found space for its animal and plant experiments. Two of its most prominent members, Auguste Hardy, of the Algiers Jardin d'essai, and Jevin Loche, director of the Algiers museum of natural history, provided the committee with space for experiments, animals and quasi-official homes. Loche, who on energetic personal initiative founded the Algiers municipal zoo, also served as the committee's delegate to the Paris society.[20]

This reliance on co-existing and in many cases pre-existing institutions, both for space and for funding, is a common feature of the history of the acclimatization societies. In South Australia, for example, George William Francis, director of the Adelaide Botanic Gardens, lectured on acclimatization with such enthusiasm that he stimulated members of the Adelaide Philosophical Society to form an acclimatization society. Francis proposed that in the event of the group's failure, all proceeds be given to the Botanic Gardens.[21] Gillbank documents a similar pattern of inter-institutional symbiosis between the Acclimatisation Society of Victoria, the Melbourne Botanic Gardens and the Zoological Gardens in Royal Park.[22]

The acclimatization society movement rippled outwards from Paris and Algiers, and although similar groups soon formed in most of Europe's major cities, none would be as large, as rich or as enduring as the Paris group. In London, a Society for the Acclimatisation of Animals, Birds, Fishes, Insects and Vegetables sprang from modest beginnings in June 1860. However, the group was ineffective in running its own show and, despite grandiose rhetoric, it proved incapable of guiding its colonial counterparts. Led by the science journalist and naturalist Frank Buckland, the group stuck close to his agenda, which was building up the game and fish preserves of England proper. A few members of the London group participated in the activities of the Paris acclimatization society, which opened its prize competitions to people of all nations. In 1862 John Bush's Chinese sheep won a silver medal at the Paris group's competition, and at about the same time Lady Dorothy Nevill completed a successful series of sericultural experiments which she supplemented with translations of French instructions on silk production. The London group never attracted more than four hundred members, and survived on its own for only five years. It might have fared better had not one of its founders, Edward Wilson, a member of the London, Paris and Victorian acclimatization societies, returned from Britain to the antipodes.[23]

In the French Empire, Parisian directives instigated the establishment of colonial acclimatization committees. This stands in stark contrast to the situation in the British Empire, where acclimatization groups arose on the British periphery in the absence of strong metropolitan incentives and leadership. No British territory boasted more voluntary acclimatization activity than Australia, where at least four separate societies functioned in the colonies of New South Wales, Queensland, South Australia and Victoria.[24] While the Queensland Acclimatization Society was also successful in terms of size and longevity,[25] Australia's first acclimatization society took form in Melbourne in 1861. The economy of Victoria, unlike that of free-trading New South Wales – a colony with a smaller acclimatization society by Australian standards – rested on protective tariffs in the early and middle 1860s.[26] In Australia, as in Algeria, the activities of acclimatization societies peaked during the last decades when economic shelters protected colonial products.

The Victorian case provides a striking parallel to that of Algeria. The economic circumstances of the two colonies were similar, and in both cases surges in activity occurred when favourable tariffs on exotic products combined with government assistance in the form of grants, land and facilities. The acclimatization societies of Algiers and Melbourne were also located in areas that underwent rapid population growth. Algeria, with little mineral wealth, attracted European settlers at a moderate rate through the 1860s. However, the city of Algiers grew more rapidly. In the antipodes, the gold rushes of the 1850s brought miners to southeastern Australia who settled after their gold fever subsided. By 1861, when the Acclimatization Society of Victoria was founded, Melbourne was a rapidly expanding industrial centre the population of which had more than quadrupled in the previous decade.[27] Only a few years later, James McPherson Grant's new Selection Act opened Victoria to a new wave of settlement.

The modality of land tenure in Victoria and Algeria also shared common features. In Algeria, the official vision was one of independent landholders settling on and developing moderate estates with the goal of self-sufficiency, and perhaps cottage industry and a few specialty crops for added income; in French terms, transplanted peasants. The Paris society captured the patronage of Napoleon III because it offered a way to implement his small-holdings settlement scheme in Algeria.[28] A philosophy of small holdings combined with exotic crop acclimatization was also used in Australia, notably by the Queensland reform advocate William Brooks.[29] A key idea linking acclimatization to land tenure was that animals like llamas were multi-functional, being at once beasts of burden and providers of fleece and meat. Llama keeping and, if tilling of the soil was required, yak keeping, made economic sense on the mental balance sheets of many acclimatizers. Acclimatization societies searched for and imported exotic multi-functional quadrupeds as a means of sparing farm families with limited resources the expense of keeping a variety of specialized animals for draught work, fleece production and meat.[30]

Abundant material exists for comparison of the theories that informed animal acclimatization experiments in French Algeria and Victoria. Intercolonial co-operation implies an exchange of objects. In the case of the French and Australian acclimatization societies, the exchange was mutual but it was not symmetric with respect to organisms and theory. Melbourne's acclimatization society took the Paris and perhaps the London society as its organizational model. Its members exchanged information and organisms with Paris and Algiers, but organisms flowed more frequently in a south to north direction as Australia became a model to be emulated for the French colonial enterprise in Algeria. Other Australian acclimatization societies, many of them sparked by Edward Wilson, looked to the Victorians for organizational design – and perhaps Hoare's phrase, 'what science has done for Victoria ... science will do for Queensland', might be applied as well to acclimatization activity in South Australia and New South Wales.[31] French zoological theories did not fare well in Australia, and most prominent Australian acclimatizers rejected the cluster of transformist ideas that animated the acclimatization movement in the French Empire. Adelaide's George Francis was just one of many who sided with Joseph Hooker's view that a real acclimatization or adaptation to new environmental circumstances was impossible.[32]

DIVERGENT PERSPECTIVES ON THE SCIENCE OF ACCLIMATIZATION

When the journalist Edward Wilson returned to Australia from London toward the end of 1860, he stimulated members of Melbourne's recently-formed Zoological Society of Victoria to reconsider the goals of their organization and to place the acclimatization of animals above their programme of zoo keeping. The resulting organization, the Acclimatisation Society of Victoria, attracted about five hundred members in the period 1861–1864 and garnered more than £ 10,000 of financial assistance from the Victorian government. Even though the group raised substantial sums of its own, its prosperity rested tenuously on government patronage. Shortly after the government withdrew its yearly matching funds of £ 1,000 zoo-keeping displaced acclimatization as the group's main function.[33] English voluntarism could be a fickle benefactor, and part of the London acclimatization society's feebleness resided in its inability to obtain government monies or attract rich patrons. George Francis realized this and argued that South Australian acclimatizers, who had no rich aristocracy to whom they could appeal, deserved government funding like the Victorians.[34]

During its heyday, the Victorian society attracted a number of notable colonial naturalists, including two men eminently qualified to speak as the group's scientific voices, Frederick McCoy and Ferdinand Mueller. Their views on the scientific nature of acclimatization differed from those current in France. Both McCoy, the University of Melbourne's professor of natural science and government palaeontologist for Victoria, and Mueller, director of the Melbourne Botanic Gardens and Victoria's government botanist, became leading members of the Acclimatisation Society of Victoria and remained vice-presidential officers until the early 1870s. They gained reputations as impassioned opponents of transformism, especially in its most accessible guise of Darwinism. McCoy and Mueller were typical of the élite of Australia's scientific community, which rejected Darwinism until the 1880s. This is not to say that an élite creationist ideology prevented the emergence of heterodox views favourable to Darwinian evolution. Two of Sydney's notable acclimatizers, George Ledger and François Laporte, accepted versions of transformism, Ledger tending toward Darwinism and Laporte accepting a wash of Lamarckian and Darwinian themes.[35] The ideas of Ledger and Laporte notwithstanding, the predominant trend among Australia's acclimatizers was to put an anti-transformist gloss on the explanation of their endeavours.

McCoy based his anti-progressionism and anti-transformism on an eighteenth-century variant of the chain of being doctrine.[36] Mueller, a devout Lutheran, rejected the idea of transmutation and declined Darwin's invitation to conduct a series of botanical experiments in Australia. Instead, Mueller declared to Richard Owen that nature always created 'well-defined unalterable species' and offered his field work in support of Owen's anti-transformist thesis.[37] McCoy's faith in the stasis of the organic world did not translate into a rejection of acclimatization, but the position led him to support only the acclimatization of animals from analogous climates.

McCoy summarized his views on acclimatization for the Acclimatisation Society of Victoria's first anniversary meeting in a lecture entitled 'Acclimatisation, its nature and applicability to Victoria'. McCoy noted that for some people acclimatization

> conveys the idea of changing the powers of enduring certain climates of different animals so as to bring animals of hot countries to live in cold ones, or *vice versa* ... But there is no such limit placed on the labours of the acclimatiser, nor does he contemplate great advantages to flow from any such attempt at changing the natures of many animals, nor does he give much trouble to this point; and, in fact, by far the greater number of important achievements of acclimatisation have been rather *the bringing together in any one country the various useful or ornamental animals of other countries having the same or nearly the same climate and general conditions of surface*. [McCoy's emphasis].[38]

McCoy's natural theological perspective found 'no authority, either in Scripture or in science, for the belief in the gradual transmutation from one species to another'.[39] For him, the process of acclimatization exploited what he called the law of representative forms, a natural consequence of specific centres of creation. Many species, although widely separated in distance, lived in 'places of similar climatal and surface characters' and hence possessed similar characteristics.[40] McCoy illustrated his lecture with a map of the globe marked by circles representing the natural habitat of several useful species. Rather than alter an animal's nature through a kind of forced adaptation, the essential task of the acclimatizer was to rearrange God's creation. Humans merely assisted useful species to pass over inhospitable climatic zones until they resettled in the corresponding latitude in a different hemisphere.

The Acclimatisation Society of Victoria followed McCoy's programme for rearranging creation, and by 1862 it had shipped animals to sister societies or related groups in London, Paris, St. Petersburg, Calcutta and Cologne. The group also purchased and assembled by exchange a fine collection of exotic ruminants, game-birds, and barnyard fowl that were housed on private estates and at Melbourne's Royal Park and Botanic Gardens.[41] The same organism that had once captured Daubenton's attention – the Spanish merino sheep – also enabled Australia to become England's major fleece supplier by mid-century.[42] Members of the Victorian society hoped to duplicate this effort by importing a new money-spinner; in 1862 their pens in Melbourne contained alpacas, llamas, Angora goats and Chinese sheep. Many members expected llamas and alpacas to produce the next major improvement in fleece production, but this would not be due to a physiological adaptation or transformation in the animals. Success seemed likely because the natural range of llamas and alpacas in Latin America was at a latitude similar to Australia's.[43]

Ferdinand Mueller also expressed his views of the scientific basis of acclimatization. Like many other curators of gardens and natural history museums, he furthered scientific co-operation beyond the British Empire.[44] Mueller exchanged useful plants with the Paris acclimatization society, the Paris Museum of Natural History, the Algiers Jardin d'essai, and institutions in many other parts of the world. A founding member of the Paris society and a member of many other European academies, Mueller also became a *correspondant* of the Paris Museum in 1862. His election was a fortunate one for the Museum professors, as Mueller added to their knowledge of Australia with shipments of plants, preserved fish and reptiles.[45] Mueller became Australia's expert on economic botany and perhaps the European scientific world's foremost authority on eucalyptus, the seeds of which he seldom tired of sending to the far reaches of the globe. Convinced of the tree's ability to dry swamps and therefore make fever-ridden areas habitable, Europeans had introduced the tree into France, Corsica, Algeria, the Iberian peninsula, Senegal, Tahiti, Egypt and many other countries by 1875.[46]

Mueller had compiled a catalogue on Australia's indigenous plants in conjunction with Melbourne's Intercolonial Exhibition of 1866–1867. His best-known publication, a manual on the economic botany of exotic plants, made its appearance in 1872 as a volume of the *Proceedings* of the Zoological and Acclimatisation Society of Victoria. Under the title, *Select Extra-tropical Plants Readily Eligible for Victorian Industrial Culture*, the book ran to at least eight general editions before Mueller died in 1896, and was supplemented by separate regional editions for Victoria, New South Wales, India and America. *Select Extra-tropical Plants* also served as the inspiration for a collaborative volume in the French language with the French botanist Charles-Victor Naudin, a member of the Paris society and former *aide-naturaliste* at the Paris Museum of Natural History.

Naudin adhered to the limited transformism common among the French acclimatizers. Like the German *Naturphilosophen*, he believed that the proliferation of the world's present organic forms had developed from a smaller number of ancestral forms through a series of transformations. These transformations or hatchings, as Naudin called them, resembled the stages an embryo experienced as it grew to adulthood. Although the great lines of organic development had been predetermined, Naudin held that many present-day species retained a measure of primordial plasticity.[47] Health problems forced Naudin from Paris to southern France, where he founded a private *Jardin d'expériences* at Collioure, Pyrénées-Orientales, before accepting the directorship of a government-funded station of agronomic research at Antibes in 1878. The Antibes research station had the oldest collection of eucalypts in France and Naudin, who botanized in Algeria in 1847, 1852 and 1883, used hybridization, artificial fertilization and other horticultural techniques to develop cash crops and medicinal plants for the colony and the south of France.[48] Intimately familiar with the plants of Australia, he authored the key scientific monograph on eucalyptus in the French language.

Naudin's affirmation of the flexibility of organic form is similar to the theory of limited variability of type promoted by Geoffroy Saint-Hilaire, at the opposite end of the spectrum from resolute fixists such as McCoy, Mueller and others who viewed acclimatization as a mere transfer of organisms between analogous climates. Naudin placed his faith in the general flexibility of plants. The techniques of horticulture, he wrote, if judiciously applied, could provoke changes and would reveal the extent of 'aptitudes hidden in the wild state'.[49] Such an approach left room for the acclimatizer to manipulate a wide range of possibilities. If Mueller and Naudin were aware of their disagreements over scientific explanations for acclimatization, they chose not to air their differences in print. The fruit of their collaboration was published in 1887 as the *Manuel de l'acclimateur ou choix de plantes recommandées pour l'agriculture, l'industrie et la médecine*.

The *Manuel de l'acclimateur*, basically a guide for the economic botany of exotic plants, assumed a revered place in the French literature of colonial development. In the volume's introduction, Naudin paid homage to Isidore Geoffroy Saint-Hilaire and the work of the Paris acclimatization society. It was the book's purpose, wrote Naudin, to provide guidance to fellow members of the Paris society and other 'numerous experimenters who have an interest in the naturalization of exotic plants, principally in Europe, North Africa, and in the French colonies'.[50] French consuls in Australia, particularly François Laporte in Melbourne, active in both the Paris and Melbourne groups, promoted the volume to the governor-general of Algeria and arranged for the shipment to Algiers of Australian plants and seeds, many of which were gathered by Mueller.[51]

Although the Paris society offered a prize for the acclimatization of kangaroos in Algeria – and there were times when French cottage industry could have used kangaroo leather for gloves – plants were easier to transport and establish in Algeria than animals.[52] The prominent place given to eucalyptus trees in manuals of colonial development such as those published by Mueller and Naudin furthered French reliance on acclimatized Australian organisms. One such publication, the *Guide du planteur d'eucalyptus*, referred to the tree as the 'tree of colonization', and promoted its use in the colony of Algeria.[53] The *Guide*, which republished many of the writings of Monsieur Ramel, an honorary member of the Melbourne acclimatization society who resided in Paris, promoted a plan suggested by Mueller to plant one quarter of Algeria's surface with trees. The author estimated that by 1876 some 1,500,000 eucalypts, many of them planted by members of the Paris society, existed in the colony. To realize Mueller's plan, ten times that number were needed, and the *Guide* gave information on how that might be accomplished.[54] Algeria's Inspector of Forests and colonial engineers coveted eucalyptus for its durability and rapid growth. Groves of eucalyptus produced a self-preserving wood suitable for telegraph poles in only five years. Pine trees, a scarce resource in colonial Algeria, also provided wood for telegraph poles. But pine trees took twenty-five years to reach telegraph pole size and needed costly treatment with preservatives.[55] By serving as railway ties, windbreaks, sanitizers of swams, and telegraph poles, the acclimatized Australian eucalyptus became a symbol and tool of French colonization in North Africa.

* * *

In considering the history of metropolitan and colonial acclimatization societies in the French and British empires, one is struck by the similarities of membership, programme and organization. A composite scenario of acclimatization society activity based on the histories of the groups active in Paris, Algiers, London, Melbourne and Sydney reveals a cast of scientists, pastoralists, government officials (usually from the diplomatic corps or a ministry with ties to the colonies) and publicists. On balance, the Paris society and the Algiers group seem to have had more scientists per capita than the British and Australian societies. In the Paris society, one member in ten was a scientist or physician; in Algeria, the ratio rose to one in six.[56] The London society, hampered by severe competition for its functions by Kew Gardens and the London Zoological

Society, attracted few scientists.[57] In Australia, where the resident population of those who could qualify as scientists was modest, acclimatization societies usually attracted a few prominent scientists. However, Edward Wilson, reflecting on the demise of the Victorian society in 1875, ascribed the success of his group to practical men and chided scientists for having remained aloof of acclimatization.[58]

It can also be said of the acclimatization societies under scrutiny that they emerged from pre-existing institutions, and often collapsed back into those same institutions. In France, the Ministry of Public Instruction encouraged this boom-and-bust pattern by fostering the proliferation of scientific societies through a programme that made new societies eligible to receive financial assistance. Bursts of government funding also assisted but then left stranded the Victorian acclimatizers, and it appears that a similar pattern of rapid institutionalization followed by moderate growth and decline occurred in the British Empire.

Comparison of the activities of the acclimatizers in Algeria and Victoria suggests two conclusions. First, French and British scientists and administrators concerned with the colonies and a substantial core of colonists collaborated to develop the non-European world. This activity, serving as it does as an example of a trans-national universalism in science, coincided with an 'anglo-philic' posture adopted by some members of the French diplomatic corps around the time of France's entry into the arena of free-trade imperialism in 1860. Second, although the French and Australian scientists who commented on the theory of acclimatization were unanimous in their rejection of Darwin's evolutionary theory, they operated from very different conceptions of how organisms related to their environment. While organisms, practical information concerning techniques, and organizational models flowed freely between Algiers and Melbourne, there seems to have been little or no true commerce in or reception of theory, specifically in accounts of what happened to the animals undergoing acclimatization or in explanations of failed experiments.

In the trans-national context, the life of these groups was tied to the development phase of the European settlement colonies. It seems too that the programme of the acclimatization societies became less essential to colonial development once a colony found its mission. In Australia, as in Algeria, the florescence of acclimatization activity occurred within a protectionist economic environment. In both colonies, interest in acclimatization fell off in the 1870s and 1880s. By the later decades of the century, economies of scale and agricultural specialization had begun to displace agricultural systems based on self-sufficiency, cottage industry, and the poly-cultural exploitation of small estates – the very situation where acclimatization societies prospered. The acclimatizing groups could still import animals for hunting and fishing, but these activities were less essential to the government and well-being of the colonies than the development of cash crops and new industries.

The administrators of Europe's preeminent colonial empires realized that much could be gained by studying the colonial development strategies of their imperial rivals. This circumstance favoured inter-colonial scientific co-operation across empires when the spheres of political influence were well-defined. In many ways transplanted Europeans saw themselves as engaged in a common battle to Europeanize the environments of their respective colonies.[59] Colonial acclimatization societies did import flora and fauna from Europe, but the predominant trend in Algeria and Australia was to look to other warm countries for source materials. Even when the selection of animals was informed by transformist zoological theories, the distinctiveness of the physical, economic and social environments of the colonies drove the acclimatization societies to appropriate the natural productions of other colonies.

Individuals such as Mueller, Naudin and Hardy played key roles in maintaining the acclimatization society networks. General MacMahon, for example, a member of the Paris society and the Algerian governor-general in the late 1860s, was a dedicated student of British colonial development who singled out Australia as an especially appropriate source of trees to plant in Algeria.[60] By the same token, the trans-shipment of flora and fauna between Australia and Algeria sprang from the conviction, held by MacMahon, Auguste Hardy, and others, that portions of both colonies fell within a 'Mediterranean' climatic zone[61] and that France could more or less appropriate what Britain was doing in Australia and then apply it to Algeria.

The political context of acclimatization society activity should not be overlooked. The French continued to use Australia as one of several models for Algerian development until late in the century. The practice drew some criticism, however, and in 1887 Louis Vignon, a staffer in the French colonies and commerce ministries, pointed to Algeria's unique characteristics and inveighed against further use of the Australian model.[62] The level of acclimatization society activity was on the wane by this time, a circumstance of competition with other cultural forums, a history of unprofitable experiments, and reductions in government funding to acclimatization societies.

For Vignon, Australia differed from Algeria on almost all counts. Australia's relatively docile native population, its temperate climate and vast tracts of arable land had enabled the colony to sustain large numbers of European settlers. Moreover, England's liberal emigration policies, coupled with the discovery of gold in 1851, had ensured sufficient supplies of capital and cheap labour. Australia, unlike Algeria, was also relatively free of fevers, a circumstance that in Vignon's mind meant that Europeans could safely perform agricultural work. Inhabiting a fever-ridden country without mineral wealth, the Algerians tenaciously resisted French rule. The French government had continued to limit emigration to the colony while labour and capital were in short supply. France, maintained Vignon, needed to relinquish the idea that Algeria was an agricultural and settlement colony like Australia. Europeans could not safely work the soil in North Africa. For Vignon, as for more famous theorists of French colonial development like Paul Leroy-Beaulieu, the best way to develop the colony was to restrict European immigration and force the natives to work on palm, sugarcane, and coffee plantations.[63]

The ascendancy of the ideas of Vignon and Leroy-Beaulieu within France contributed to the decline of France's reliance on Australian organisms in the development of Algeria and of acclimatization activity in the colonies. As Australian and Algerian agriculture emerged from the cottage industry stage, the strategy for development shifted and left behind the acclimatization projects of the 1860s and 1870s. However, the history of the acclimatization societies in the European colonies, bound as it is to a certain stage of colonial development, provides a point of entry to the comparative analysis of colonial science, and the diffusion (or in this case the non-diffusion) of scientific theory across empires.

Department of History
University of California, Santa Barbara
U.S.A.

NOTES

* I wish to thank Linden Gillbank who unselfishly shared her knowledge of the Acclimatisation Society of Victoria with me. Sally Gregory Kohlstedt, William Coleman, Anita Guerrini and the staff of the Mitchell Library in Sydney also provided

invaluable assistance. This research was supported by grants from the Wisconsin Alumni Research Foundation, the American Council of Learned Societies, and the Centre National de la Recherche Scientifique.

1 Mueller to E.A. Pesoli, 5 October 1883. I thank Professor Wallace Kirsop of the Department of Romance Languages, Monash University, for providing me with a copy of this letter which will appear in the edition of Mueller's correspondence under preparation by Professor R.W. Home and his colleagues.

2 Roy MacLeod, 'On visiting the 'moving metropolis': Reflections on the architecture of imperial science', *Historical Records of Australian Science*, 5(3) (1982), 1–16; reprinted in *Scientific Colonialism: A Cross-Cultural Comparison*', ed. N. Reingold and M. Rothenberg (Washington, D.C., 1987), pp. 217–249. Idem, 'Introduction', *The Commonwealth of Science: ANZAAS and the Scientific Enterprise in Australasia, 1888–1988*, ed. R. MacLeod (Melbourne, 1988), pp. 1–16; p. 5.

3 Joseph M. Powell, *Environmental Management in Australia, 1788–1914: Guardians, Improvers and Profit, an Introductory Survey* (Melbourne, 1976); idem, *Mirrors of the New World: Images and Image-Makers in the Settlement Process* (Kent, UK, 1977). Susan Sheets-Pyenson, 'Cathedrals of science: The development of colonial natural history museums during the late nineteenth century', *History of Science*, 25 (1987), 279–300; idem, 'Civilizing by nature's example: The development of colonial museums of natural history, 1850–1900', in *Scientific Colonialism*, op. cit. (n. 2), pp. 351–377. Her book, *Cathedrals of Science*, was unavailable to me before finishing this essay. Sally Gregory Kohlstedt, 'International exchange and national style: A view of natural history museums in the United States, 1850–1900', in *Scientific Colonialism*, op. cit. (n. 2) pp. 167–190. Lewis Pyenson, *Cultural Imperialism and the Exact Sciences: German Expansion Overseas, 1900–1930* (New York, 1985); idem, 'The limits of scientific condominium: Geophysics in Western Samoa, 1914–1940', *Scientific Colonialism*, op. cit. (n. 2), pp. 251–295. Alfred W. Crosby, *Ecological Imperialism: The Biological Expansion of Europe, 900–1900* (Cambridge, 1986).

4 Kew Gardens directed much of the British Empire's plant acclimatization activity. See Lucile H. Brockway, *Science and Colonial Expansion: The Role of the British Royal Botanic Gardens* (New York, 1979). Daniel R. Headrick's valuable account of plantation agriculture and tropical botany in the British, Dutch and French colonies leaves out Australia and Algeria. See his *The Tentacles of Progress: Technology Transfer in the Age of Imperialism, 1850–1940* (New York, 1988), pp. 209–258.

5 John Sutherland et al., *Report of the causes of reduced Mortality in the French Army Serving in Algeria, presented to both Houses of Parliament* ... (London, 1867).

6 Recent case studies of French scientific societies in the post-Revolutionary era which include colonial themes are Alfred Fierro, *La Société de Géographie, 1821–1946* (Geneva, 1983); and Michael A. Osborne, 'The *Société zoologique d'acclimatation* and the New French Empire: The Science and Political Economy of Economic Zoology During the Second Empire' (Ph.D. dissertation, University of Wisconsin-Madison, 1987). Case studies by Robert Fox include, 'La Société zoologique de France: Ses origins et ses premières années', *Bulletin de la société zoologique de France*, 101(5) (1976), 799–813 and 'Presidential address: Science, industry, and the social order in Mulhouse, 1798–1871', *British Journal for the*

History of Science, 17(1984), 127–168. General studies by Fox are 'The *savant* confronts his peers: Scientific societies in France', in *The Organization of Science and Technology in France, 1798–1871*, ed. R. Fox and G. Weisz (Cambridge, 1980), pp. 241–282 and 'Learning, politics and polite culture in provincial France: The *Sociétés savantes* in the nineteenth century', *Historical Reflections/Réflections historiques*, 7(2/3) (1980), 543–564. French scientific societies are featured in Harry W. Paul, *From Knowledge to Power: The Rise of the Science Empire in France, 1860–1939* (Cambridge, 1985), pp. 251–287. A perspective on British societies is *Metropolis and Province: Science in British Culture, 1780–1850*, ed. Ian Inkster and Jack Morrell (Philadelphia, 1983). Enlightenment scientific societies form the theme of James E. McClellan III, *Science Reorganized: Scientific Societies in the Eighteenth Century* (New York, 1985).

[7] Linden Rae Gillbank, 'The Acclimatisation Society of Victoria', *Victorian Historical Journal*, 51 (1980), 255–270; idem, 'The origins of the Acclimatisation Society of Victoria: Practical science in the wake of the gold rush', *Historical Records of Australian Science*, 6 (1986), 359–374. The general context of Australian scientific societies is given in Michael Hoare, 'Science and Scientific Associations in Eastern Australia, 1820–1890' (Ph.D. dissertation, Australian National University, 1974). Useful too are his 'Botany and society in Eastern Australia', in *People and Plants in Australia*, ed. D.J. and S.G.M. Carr (Sydney, 1981), pp. 183–219 and 'The intercolonial science movement in Australia', *Records of the Australian Academy of Science*, 3(2) (1976), 7–28.

[8] *Annual Report of the Acclimatisation Society of New South Wales, 1861* (Sydney, 1862), p. 7.

[9] Osborne, op. cit. (n. 6), pp. 72–82, 'Charles de Montigny, naturalist-diplomat'.

[10] Ibid., pp. 208–210; F.E. Guerin-Méneville, *Rapport à S.M. l'empereur sur les travaux entrepris par ses ordres pour introduire le ver à soie de l'aylanthe en France et en Algérie* (Paris, 1860).

[11] Mitchell Library, Sydney. Isidore Geoffroy Saint-Hilaire to Sir William Macarthur, 22 December 1855, Macarthur Papers, Vol. 40; linotype from Société zoologique d'acclimatation, 28 February 1861, and Alfred Roberts to Sir William, 5 November 1860, Macarthur Papers, Vol. 41. For biographical details see Ruth Teale, 'Sir William Macarthur', in *Australian Dictionary of Biography*, Vol. 5, ed. D. Pike (Melbourne, 1974), pp. 124–125.

[12] E. Roehn and B. LaPommeraye, *Mémoire sur l'introduction en France de la race des alpacas et llamas ...* (Paris, 1848); Osborne, op. cit., n. 6, p. 91. Most antipodean llamas and alpacas came from an expedition mounted by C. Ledger. See C. Ledger, 'Paper on the Alpacas', *Sydney Morning Herald*, 2 February 1864.

[13] George Bennett, *Acclimatisation: Its Eminent Adaptation to Australia: A Lecture Delivered in Sydney Republished by the Acclimatisation Society of Victoria* (Melbourne, 1862), p. 40. For Bennett's relationship with Richard Owen and the positions the two took on the transformism of Etienne Geoffroy Saint-Hilaire, see the chapter by Elizabeth Dalton Newland in this volume.

[14] Osborne, op. cit. (n. 6), pp. 62–128, 259–273.

[15] *First Annual Report of the Acclimatisation Society of Victoria* (Melbourne, 1862), p. 7; *Annual Report of the Acclimatisation Society of New South Wales, 1861*,

(Sydney, 1862), p. 7.
16 Richard du Cantal, *Titres et travaux scientifiques de M. Richard (du Cantal)* (Paris, 1868), p. 4.
17 R. du Cantal, 'Rapport ... sur les espèces chevaline, asine, bovine et porcine de l'Algérie', *Bulletin de la Société d'acclimatation*, 4 (1857), 303–330, 365–390.
18 Ledger, op. cit. (n. 12), and 'Sur les alpacas et sur les avantages que présente leur introduction en Australie', *Bulletin de la Société d'acclimatation*, 8 (1861), 33–37. *Sixth Annual Report of the Acclimatisation Society of Victoria* (Melbourne, 1868), p. 6. Acclimatization societies often rationalized pre-existing activities. The Angora goat herd pre-dated the formation of the Acclimatisation Society of Victoria; see Gillbank, 1986, op. cit. (n. 7), p. 370.
19 The *Sixth Annual Report of the Acclimatisation Society of Victoria* (Melbourne, 1868), pp. 20–21, indicates that the preceived profitability of ostrich farms in Algeria stimulated antipodean interest. Gillbank, op. cit. (n. 7), notes too that the Melbourne Zoological Society was engaged in acclimatization before it transformed into an acclimatization society. Acclimatization societies were also concerned with the efficient exploitation of native resources; see François Laporte's review of Australian ichthyology in *Proceedings of the Zoological and Acclimatisation Society of Victoria and Report of the Annual meeting of the Society*, Vol. 1 (Melbourne, 1872), pp. 28–248; and the 'Report on the ostriches' in Vol. 2 (Melbourne, 1873), pp. 205–232 in the same series.
20 Osborne, op. cit. (n. 6), pp. 192–257.
21 George W. Francis, *The Acclimatisation of Harmless, Useful, Interesting, and Ornamental Animals and Plants*. Being a paper read before the Philosophical Society, Adelaide, South Australia on May 13th 1862 (Adelaide, 1862) p. 22.
22 Gillbank, 1986, op. cit. (n. 7).
23 Osborne, op. cit. (n. 6), pp. 347–364; Gillbank, ibid., pp. 362–374.
24 Eric C. Rolls, *They All Ran Wild* (Sydney, 1969), pp. 209–275; C.F.H. Jenkins, *The Noah's Ark Syndrome: One Hundred Years of Acclimatization and Zoo Development in Australia* (Perth, 1977). The Mitchell Library catalogue has entries for sixteen Australasian acclimatization societies.
25 Hoare, 1974, op. cit. (n. 7) p. 240; and 1981, op. cit. (n. 7), p. 207.
26 A.J. Reitsma, *Trade Protection in Australia* (Leiden, 1960).
27 David Dunstan, *Governing the Metropolis: Politics, Technology and Social Change in a Victorian City, 1850–1891* (Melbourne, 1984).
28 Osborne, op. cit. (n. 6), pp. 134–137, 225–233.
29 Hoare, 1974, op. cit. (n. 7), pp. 241–242.
30 The relationship between multi-functional animals such as yaks, the division of labour, and peasant agriculture is considered in A. de Quatrefages de Bréau, 'Notice sur les yaks et les chèvres d'Angora importés en France depuis la foundation de la société', *Bulletin de la Société zoologique d'acclimatation*, 4 (1857), li–lx; pp. liv–lv.
31 Hoare, op. cit. (n. 7), p. 228.
32 Francis, op. cit. (n. 21), p. 5.
33 Gillbank, 1980, op. cit. (n. 7), p. 266; Rolls, op. cit. (n. 24), p. 272.
34 Francis, op. cit. (n. 21), p. 9.
35 Barry W. Butcher, 'Darwin's Australian Correspondents: Deference and Collabora-

tion in Colonial Science', in *Nature in Its Greatest Extent: Western Science in the Pacific*, ed. Roy MacLeod and Philip F. Rehbock (Honolulu, 1988), pp. 139–157; Ledger, op. cit. (n. 12); Laporte, op. cit. (n. 19), p. 39.

[36] Ann Mozley Moyal, ed., *Scientists in Nineteenth Century Australia: A Documentary History* (Sydney, 1976), pp. 186–201; Gillbank, op. cit. (n. 7), p. 258.

[37] Mueller to Richard Owen, in Moyal, op. cit. (n. 36), p. 193.

[38] Frederick McCoy, 'Acclimatisation, its nature and applicability to Victoria', in *Acclimatisation Society of Victoria, First Annual Report* (Melbourne, 1862), pp. 33–51; p. 36.

[39] Professor [Frederick] McCoy, *The Order and Plan of Creation: Two Lectures Delivered in Connection With the Early Closing Association* (Melbourne, 1870), p. 10.

[40] McCoy, 1862, op. cit. (n. 38), p. 37.

[41] Ibid., pp. 42–43.

[42] Philip McMichael, *Settlers and the Agrarian Question: Foundations of Capitalism in Colonial Australia* (Cambridge, 1984), pp. 54–75.

[43] McCoy, 1862, op. cit. (n. 38), pp. 42–43, 49–50.

[44] Kohlstedt, op. cit. (n. 3).

[45] Meetings of the Museum professors, 29 April, 13 May 1862, F^{17} 3883, Archives nationales, Paris.

[46] E. Cosson, 'Note sur l'acclimatation de l'*Eucalyptus Globulus*', *Extrait du Bulletin de la Société de géographie* (juin 1875) (Paris, n.d.).

[47] John Farley, 'The Initial reactions of French biologists to Darwin's *Origin of Species*', *Journal of the History of Biology*, 7 (1974), 275–300.

[48] Berthelot, *Notice historique sur la vie et les travaux de M. Naudin* (Paris, 1900).

[49] Charles-Victor Naudin and Ferdinand von Mueller, *Manuel de l'acclimateur ou choix de plantes recommandées pour l'agriculture, l'industrie et la médecine* (Antibes, 1887), p. 5. Divergent perspectives on Isidore Geoffroy Saint-Hilaire's transformism are Goulven Laurent, *Paléontologie et évolution en France de 1800 à 1860: Une histoire des idées de Cuvier et Lamarck à Darwin* (Paris, 1987), pp. 467–489; and Osborne, op. cit. (n. 6), pp. 259–273.

[50] Naudin and Mueller, op. cit. (n. 49), p. 2.

[51] See for example MacMahon to M. le Ministre des Affaires Etrangères, 30 April 1868; F. de Castelnau [Laporte] to Governeur-général de l'Algérie, 21 January 1878; Consul de Melbourne to Governeur-général de l'Algérie, 24 December 1888, Affairs diverses commerciales, carton 222bis, 'Algérie 1861–1893', archives of the Ministère des Affaires Etrangères, Paris.

[52] Osborne, op. cit. (n. 6), p. 105; Bennett, op. cit. (n. 13), p. 6.

[53] A. Certeux, *Guide du planteur, d'eucalyptus* (Algiers, 1877), p. 13.

[54] Ibid., pp. 14–18. For Mueller's plan, first published in the 1869 volume of the *Bulletin de la société d'agriculture d'Alger*, see p. 146.

[55] Edward Pepper, 'Eucalyptus in Algeria and Tunisia from an hygienic and climatological point of view', *Proceedings of the American Philosophical Society* 35 (1896), 39–56; Certeux, op. cit. (n. 53), pp. 150–152.

[56] Osborne, op. cit. (n. 6), pp. 64, 116, 213–221.

[57] Ibid., pp. 356–358.

58 Edward Wilson, *Acclimatisation*, Read before the Royal Colonial Institute (London, 1875), p. 22.
59 Crosby, op. cit. (n. 3).
60 MacMahon, op. cit. (n. 51).
61 L. Emberger, 'La place de l'Australie méditerranéenne dans l'ensemble des pays méditerranéens du vieux monde', in *Biogeography and Ecology in Australia*, ed. A. Keast et al. (Den Haag, 1959), pp. 259–273.
62 Louis Vignon, *La France dans l'Afrique du Nord* (Paris, 1887), pp. 7–13.
63 Paul Leroy-Beaulieu, *De la colonization chez les peuples modernes*, 5th ed. rev. (Paris, 1902), p. 371.

SALLY GREGORY KOHLSTEDT

INTERNATIONAL EXCHANGE IN THE NATURAL HISTORY ENTERPRISE:

Museums in Australia and the United States

The Colonists settled in the antipodes were initially on the intellectual as well as the geographical fringes of natural history research. They were, nonetheless, important components in a system whose major connections reached vertically from colonizer to colonized and, to a lesser extent, horizontally throughout the colonies and former colonies.[1] Both American and Australian naturalists were, to European and British thinking in the late eighteenth and early nineteenth centuries, as essential for providing animals, minerals, rocks and plants as they were peripheral to discussions of the taxonomic and theoretical meaning of new discoveries.

World-wide efforts established comprehensive (although tantalizingly incomplete) catalogues of plants, animals, rocks and minerals. The data contributed to interpretations of geological time, topographical and climatic shifts, and world ecologies. During the nineteenth century museums preserved proof specimens and described their holdings in catalogues and journals. Outside, amateurs were asked to gather specimens, describe them, pack them carefully and transmit them to more learned colleagues. In this informal system a few colonists became well connected to the centres of intellectual inquiry and enjoyed status at home and abroad by virtue of their long-distance affiliation with noted authorities. Colonial museums, too, echoed global initiatives even as their management reflected more local public process. Dispersed curators had, at least initially, few incentives to link up with other colonials. The initial and intimate connection between Britain and her individual colonies influenced and often impeded the relationship between the United States and Australia, particularly during the troubled decades following the American Revolution. In Britain, for example, members of *national* learned societies, whose fraternity automatically intersected with politicians, merchants in international trade, and civil servants, took on the task of describing and defining the topography, plant and animal forms, and human inhabitants of the entire *world* in scientific terms.

R.W. Home and S.G. Kohlstedt (eds.), International Science and National Scientific Identity, 121–149.

Commercial and diplomatic goals inevitably became intertwined with intellectual ones in the sometimes collaborative and often competitive efforts of explorers representing different countries.

Yet there were connections made. The fact that citizens in the nineteenth century and historians in the twentieth century have recognized several significant geographical and political similarities and instances of intensive interaction between Australia and the United States raises the issue of their relationship in less well-studied intellectual and cultural matters.[2] The existence of a frontier and the trans-Pacific gold rushes in the 1840s and 1850s have been analysed comparatively. The results suggest a mutual self-consciousness among at least certain groups of Americans and Australians. Somewhat later, in 1872, Anthony Trollope commented after a visit to Australia: 'Where there is any divergence from the old John Bull proclivities [among Australians], it is toward the American side; but the divergence is not great ... '[3] Trollope's observation, with its qualifier, suggests a sense of shared identity that would grow even stronger after the federation of Australia in 1901. Still, this realignment of affinities noted by Trollope was gradual and sporadic, spurred in the twentieth century by the assertive Americans and affected by the relative decline of Britain's overseas empire. This paper explores one aspect of internationalism by investigating natural history museums in the two predominantly English-settled sets of colonies that became the United States and Australia, as they came to operate in closer view of one another.

It would be difficult if not impossible to quantify the growing communication among naturalists in Australia and United States by searching out all private correspondence, identifying every trans-national publication, or counting every object exchanged. Scientific institutions, however, with their extensive records and commitment to collective action, provide a way to trace out, over time, a qualitative pattern of activity that reveals prevailing attitudes, practical connections and, perhaps inevitably, competition. Natural history museums have been, in modern western societies, the principal location for the collection, preservation and study of natural objects. Once well established, they become the repository for materials acquired by government agencies, the disseminators of information about local animals, plants, minerals and topography, and the chief agency through which specimens and publications are exchanged. Their reputation depends not only on the strength of their holdings but also on such reciprocal relations.

Arrangements for the exchange of information and ideas, over time and with particular reference to the United States and Australia, reveal something of the redistribution of talent, the reformulation of intellectual authority, and the realignment of national interests that has occurred during the last two centuries. In the eighteenth and early nineteenth centuries, questions about distribution, identification and classification required data, acquired primarily by explorers and colonists, that were subsequently analyzed in Stockholm, London and other major urban centres and universities in Europe. In the late twentieth century Australian museums and universities are part of an international network, with strong ties not only to Britain and Europe but also to Asian and Pacific-rim countries including the United States.

Natural history, as it came to be defined and managed by museum curators, was part of the negotiation, exchange, collaboration and competition initiated between Australia and the United States. The two countries have many similarities but also vastly disproportionate resources.[4] In fact, it is precisely the difference in resources and orientation that seems most obvious in comparing the structure of museums in the United States, where private individuals and groups took primary initiative and responsibility, and Australia, where government legislation and funding were fundamental. Over time, however, the patterns of activity and support have coincided to a greater and greater extent.

AMERICANS ON AUSTRALIA

Virtually no significant natural history collections pre-dated the Revolution in the British colonies of North America. In the early years of nation building, private individuals, learned societies, proprietors and colleges established collections. None was sponsored directly by any government, although space in public buildings and donation of specimens acquired on surveys were fairly common contributions 'in kind'. Charles Willson Peale was granted space in the former national capital (then the Old State House) in Philadelphia while Samuel Scudder displayed his collection in public buildings owned by New York City in the 1810s, for example, but neither was ever able to secure direct subsidies.

Over the next thirty years cabinets, collections, and museums (terms with overlapping meanings in the nineteenth century) relied on private sponsors and active exchange. Friends, members, college alumni and patrons – especially captains of merchant sailing ships, missionaries and other world travellers – proved generous with donations mailed or carried back from throughout the Pacific Basin and Asia. Such informal patronage worked well, and by the middle of the century every major city boasted a proprietary or society museum containing local natural history specimens, displayed along with objects from Asia and occasionally a few from Australia.[5]

Rarities from abroad were not the central feature of American museums. Most of the shells, minerals, stuffed birds and mammals, pinned insects and pressed plants were familiar types gathered locally in an effort to document the local landscape. These regionally distinctive materials attracted general tourists and those with special interests in natural history. They also became a kind of 'currency' used to acquire other specimens from more distant places, as the extensive exchange and accession books of learned societies make clear.[6] The standards for collection building were broadly European, and Americans moved during the course of the century towards distinctive public and educational functions essential for making their institutions economically self-sustaining. Programmes for general audiences, for school teachers and for pupils became standard components, in part because such activities gained support from local legislatures. By 1900, museums depended on a combination of membership, philanthropic and government support.

Relatively few Americans were directly involved in the systematic exploration of the Australian continent until the twentieth century, although specimens and artifacts from the South Pacific and eastern Asia gradually accumulated in the North American collections.[7] Charles Peale, for example, displayed a lyre-bird and a duck-billed platypus that had been acquired not directly from Australia but through a London dealer. Some museums, such as the Museum of the East India Marine Society (now part of the Peabody Museum) and the Essex Institute in Salem, Massachusetts, based their holdings on objects brought back by captains of whalers and commercial ships.[8] Natural specimens were often perishable. More common, therefore, were native-made materials – carved masks, delicate inlaid boxes, decorated weapons, wooden carvings, embellished shells – as often displayed in private homes as donated to local learned institutions. The few stuffed specimens of Australian kangaroos or brilliantly colored birds, as exotic as African elephants and tigers, reminded Americans that in the antipodes, many things were 'queer and opposite'.

When the Wilkes Expedition used Port Jackson as a rendezvous and base for refueling and repairing its ships, its naturalists gained first-hand knowledge of Australia and established contact between American and Australian naturalists. In 1839, Charles Wilkes and his crew were warmly welcomed by Australians, and at the same time the first American consul, James H. Williams, took up his post in Sydney. Neither event, however, signalled a significant shift in Australian-American relations. Rev. William Clarke, then among the most active Australian naturalists, assisted the United States naturalists, including botanist William Rich, zoologist Titian Ramsey Paele and geologist James Dwight Dana. Much of the expedition material was eventually housed in the Smithsonian Institution, although some specimens went to institutions whose staff had helped finalize the multi-volume report.[9]

The Academy of Natural Sciences of Philadelphia, under the patronage of Dr. T.B. Wilson, had the largest and most comprehensive bird collection in the United States. While in London in 1847, Wilson's brother arranged to purchase Gould's famous collection of over 1,800 Australian bird skins, including type specimens, for $ 5,000.[10] Just a decade later, in 1857, Gould visited Philadelphia and other parts of the United States in order to study hummingbirds and to publicize his illustrated books on ornithology. By that time the Academy had sufficient specimens from around the world that it indicated geographical distribution using colour-coded labels for continents – Australia was assigned green. Wilson's munificence was the first in a century-long tradition by philanthropists that created specialized and named collections in museums.

Museum curators in the late 1850s relied heavily on direct and systematic institutional exchange to expand their holdings. On the American side, the ambitious Louis Agassiz of the Museum of Comparative Zoology at Harvard and the collaborative Spencer F. Baird at the Smithsonian Institution established contacts with Australian museums that would be ongoing, if intermittent.

The Swiss-born Louis Agassiz, in particular, developed an extraordinary network of correspondents as he acquired specimens from throughout the world using expeditions, exchange and purchase.[11] Australia soon came within his purview as well. In 1860 he inquired about the possibility of sending a student to explore the coasts of Tasmania, requesting government support for a knowledgeable collector and promising in return to give all duplicates to the Royal Society in Hobart.[12] Although that project apparently did not materialize, Agassiz had better success in his contacts with Ferdinand Mueller and Frederick McCoy of Melbourne. From the former he received a set of fossils and with the latter he opened a series of exchanges that enriched the holdings of the Museum of Comparative Zoology as well as the National Museum of Victoria. In 1866, for example, McCoy sent a large zinc-lined box with 33 mammals, 119 birds, 3 reptiles and 2 fish plus other items, making a total of 157 specimens and 117 species that arrived in 'excellent condition'.[13] Not content, however, with the 'common things' acquired through McCoy, Agassiz's son and successor at the museum, Alexander Agassiz, also contacted local dealers in order to acquire a more varied selection of Australian specimens.[14]

The most important long-term conduit for national and international exchange was the Smithsonian Institution, whose involvement signalled an important shift towards government sponsorship of natural history collections. Under Secretary Joseph Henry, the Smithsonian committed itself to the dissemination of information and developed an ever-expanding network for the distribution primarily of scientific publications, which eventually extended under Assistant Secretary Spencer F. Baird to specimens as well. In 1860, McCoy wrote to Henry to point out that his National Museum of Victoria, housed at the University of Melbourne, was 'one of the important Scientific Institutions omitted' from the Smithsonian's published list of foreign correspondents, and that he as director was 'actively engaged in the investigation & preparation for publication of the recent and fossil zoology of Victoria'.[15] McCoy eventually served as a link between institutions in Victoria (and Tasmania) and the Smithsonian's network of Americans.[16] Most often the specimens were exchanged for display purposes, although occasionally the data were used to pursue taxonomic research. Louis Agassiz, in fact, hoped to do original work on Australian fishes, and paleontologist O.C. Marsh sought fossils that might inform his own work on prehistoric types.[17]

Communication was often slow, expensive and difficult. By 1876, however, regular steamship routes permitted Americans with a taste and pocketbook for travel to embark on a direct line across the Pacific to the South Sea Islands and thence to New Zealand and Australia.[18] By then, too, Australians were eager to send demonstrations of their continental resources to the Centennial Exposition. When its new museum formally opened in 1881, the United States National Museum staff could claim a 'tolerably complete' set of Australian birds and initiate exchange for a more extensive study collection of marsupials as well.[19]

Other museums acquired access to Australian materials through America's most active and respected dealer in natural history specimens, Henry A. Ward, who had shipped a display of American specimens to New Zealand and then to the Melbourne International Exposition in 1881.[20] As a result of this trip, the entrepreneur sold and exchanged a significant proportion of his American materials and on a four-month tour acquired eighty boxes of Australian 'birds and beasts' to be distributed in the United States.[21] Literate Americans could also read about the 'fossil continent' in national magazines and visit popular and peripatetic panoramas and circuses that highlighted life in the 'United States of the Pacific'.[22]

In 1896 Henry Ward returned to Australia, having been hired by Alexander Agassiz to collect coral, echinoderms and sponges along the Great Barrier Reef, an effort that may have marked the last significant effort to gather Australian materials without much co-operation from Australian naturalists.[23] Agassiz had preceded Ward in April and May of that year, intending to study the developmental patterns of the coral reefs as he had all the major reef systems of the Atlantic and some of the Pacific Ocean. The oceanographer was convinced that Dana's and Darwin's theories about subsidence were oversimplified and incorrect. Unfortunately he rented a steamer at the height of the trade wind season and in one month of cruising was able to spend only three days on the outer reefs; his Australian observations where therefore inconclusive.[24] His acquisitions and theories, however, caught the attention of some Australian naturalists who evidently resented the initiative of this outsider and rejected his offer to participate in expeditions to Funafuti to study coral formations there.[25] The museum curators were subsequently in the vanguard of efforts to see that scientific work by Americans in Australian territory would be in some way collaborative with Australians.

AUSTRALIANS ON AMERICA

Early museum developments in Australia depended almost entirely on colonial government sponsorship, including direct subsidies and donations of specimens from other agencies.[26] Although the British encouraged colonists to establish natural history collections in order to identify their material resources, the mother country provided primarily moral support. Various personal incentives were sufficient to encourage individual colonial officers to establish collections that later became the core of government-sponsored museums or, in some cases, the basis of private cabinets. Australian governors and legislators, hard pressed by other financial demands, provided only the essential support required for maintenance and display of the public collections – and not always that.

Perhaps many early colonists agreed with Edward Smith Hall, who editorialized in the *Sydney Monitor* that if locals wanted to support a colonial botanist, then 'let our rich men promote it by private subscription' and if 'the old rich countries want local information in science let them send their travellers to us'.[27] That attitude sustained Australia's marginality. While travellers came and took away rare and unclassified specimens, some local naturalists sought ways to provide in the colonies some 'scientific arrangement ... toward the cultivation and domestication of the indigenous plants and shrubs of this country' as well as the animal and mineral resources.[28] Financial constraints as well as a lack of scientific expertise and reference books constrained their enthusiasm.

The scientific staff for field surveys and curatorial work at Australian museums would, until the 1880s, come almost exclusively from abroad, principally from Britain and Germany. Ambitious explorers, surveyors and curators arrived in the young colonies with high ambitions but found it difficult to sustain local meetings, publish scientific proceedings, sponsor expeditions or conduct exchanges – activities taken for granted 'back home'. Museum administrators who pressed specific agendas too hard found themselves isolated or under attack.[29] Because the Australian museums lacked authority and sustained support, they did not, in fact, control the flow of Australian materials abroad, much of which went on privately and through government channels to England or European countries. The Americans figured only marginally in either informal or formal communication throughout most of the nineteenth century. Even the American-born Alexander Morton, who immigrated to Sydney, worked at the Australian Museum, and eventually directed the Royal Society's museum in Tasmania, seems not to have maintained any special links back to the United States.

William Sharp Macleay, who visited the United States in 1836 while serving a judicial appointment in Cuba, may have been the only Australian naturalist to have visited the United States in the first half of the nineteenth century. Well-received in Philadelphia, the 'philosophical naturalist' returned to New South Wales, where his father Alexander McLeay (as the older man spelled his name) was a leading member of the public museum's oversight committee. The two, buttressed by the connections of the well-placed Macleay family, developed extensive private collections that were, in early years, superior in many ways to that of the Australian Museum.[30] The implicit competition of their holdings (usually viewed together with that of W.S. Macleay's cousin William J. Macleay), as well as paltry governmental support, frustrated a succession of curators at the Australian Museum in Sydney, one of whom was fired after he implied that William J. Macleay had increased his private collection through connection with the public institution.[31]

In 1849 the Australian Museum moved to its own building, and by the late 1850s, Sydney's Australian Museum, the Tasmanian Royal Society's museum and Melbourne's National Museum were suitably housed, open to the public, and ready to enlarge the scope of their collections through exchange. The timing fortuitously coincided with the stabilizing of publication efforts in learned societies in the United States and the creation of the Smithsonian Institution's international exchange service for publications. From this point on there would be more scientific communication between Britain's former colonies and an increasingly independent set of newer ones in the southern hemisphere.

In 1856 William S. Macleay, as a trustee at the Australian Museum, drew up a list of museums to which a form letter inviting exchanges might be sent, introducing a tactic for accession already common among European and American museums. He included 16 British, 31 European, 11 colonial and 4 American institutions, the latter being the Lyceum of Natural History of New York, the Academy of Natural Sciences of Philadelphia, the Philosophical Society of Washington [perhaps he meant the essentially defunct National Institute], and the University of South Carolina.[32] The Governor General was pleased with the number of replies, but the museum's oversight committee soon realized that they were ill equipped to answer the rapid and enthusiastic responses to their invitation for exchange. The curator reported to the trustees that the Australian Museum lacked a catalogue and sufficient duplicates to 'answer the demands of various foreign institutions and individuals' wishing to make exchanges with the Museum.[33] It was difficult to accumulate and preserve even fairly common Australian materials, because curators had limited time for exploring expeditions and little money to hire one of several independent natural history collectors who also serviced foreign buyers.[34] The extensive exchanges envisioned by Macleay never developed and, with some notable exceptions, collections of non-Australian materials remained small.

Transport was a problem for Australians as well as Americans. Shipping costs were high, security and preservation were difficult to ensure, and the service was unreliable. The Queensland Museum, for example, could only send materials to the eastern United States via London in spring and summer.[35]

As in political and economic matters, Americans figured relatively little in the routine activities of collection-building and exchange among nineteenth-century Australian naturalists. The Wilkes Expedition in 1839 led to surprisingly few sustained contacts between American and Australian naturalists and those made by Agassiz and the Smithsonian staff were intermittent. Australians remained connected by political philosophy, institutional affiliation and intellectual practices to England, with the result that their specimens, publications and scientific results went first to the mother country which returned, increasingly to their irritation, relatively little. A significant proportion was carried back to Germany by visitors and paid collectors.[36] Australians most identified Americans with economic initiatives after American miners joined the various gold rushes in Victoria and South Australia, American investors developed railroad systems, and entrepreneur S.A. Rowe took his American circus Down Under.[37]

Among the museum administrators, Frederick McCoy of the National Museum of Victoria was probably the most eager to exchange with North Americans. His own theory about the distribution of animal forms, not too far removed from Agassiz's ideas of separate creations, led him to museum displays based on geographical distribution. The preliminary list of specimens held by the Melbourne museum in 1858 contained no mention of North American species.[38] That lacuna was remedied by McCoy's subsequent exchange with the Museum of Comparative Zoology and the Smithsonian Institution. McCoy especially sought specimens exclusive to North America and appropriate to his interest in geographic distinctions among species. From Agassiz he acquired elk, deer, and a rocky mountain goat, while Baird provided fishes taken by the US Fish Commission along American shores and identified by David Starr Jordan and Pierre L. Jouy.[39] By the time the International Exposition was held in Marvellous Melbourne in 1881, that city had perhaps the largest single display of American animals in Australia, one that was enlarged even further by purchases from the natural history dealer Henry A. Ward.[40] What McCoy could not get by exchange he purchased with funds from an ever more reluctant state legislature. Special access to rare materials proved an advantage to the financially pressed South Australian Institute Museum directed by Frederick George Waterhouse. He found that aboriginal artifacts, while difficult to obtain, brought a quite substantial exchange, including the plaster cast of the skeleton of the megatherium from Ward's Scientific Establishment.[41]

The situation differed to some extent between colonies, based on geography and the outlook of the director and trustees, as well as intrinsic resources. Yet all of them benefitted from the Smithsonian's policies. The publications exchange system by 1874 had twenty-seven corresponding institutions in Australia and thirteen in New Zealand to which the Smithsonian sent its own publications, some from other American societies, and even occasional specimens.[42] The Smithsonian Institution thus played an important part in making American research available to Australians. These materials, essentially free, were especially important to the financially pressed Australian museums, which had difficulty maintaining subscriptions to British and other overseas journals and buying books.[43] In fact, the 'liberal donations of books from the American Government, the Smithsonian Institution, and other institutions and individuals in America' led the Tasmanian Royal Society to prepare a special case in the library for their reception and exhibition.[44] In Queensland the curator reported on all donations to the library and added, 'but the most numerous examples of this liberality have, as usual, been received from the United States through the Smithsonian Institution'.[45] In return the Smithsonian received the irregular publications of Australian museums and their related societies. Private collectors also occasionally used the Smithsonian conduit, in part because sending materials overland via public transport was prohibitively expensive and in part because the government agency provided a more secure system.[46] The Smithsonian also distributed sets of 'duplicates' acquired by various western exploring expeditions. While the majority went to the growing number of urban and college museums in the United States, some also went abroad. In the 1860s, for example, in Canterbury Museum in Christchurch and the Colonial Museum in Wellington, New Zealand, both received sets of muridae (rodents), undoubtedly for comparison with the imported pests on their islands.

After the 1876 exhibition in Philadelphia, there was increased contact as Australian museums gained stability and staff. To some extent this was fostered by the growing effort to make museums 'interesting and instructive' to a general public by an increasingly education-conscious group of museum administrators.[47] The commissioner of the exhibition display from New South Wales offered to exchange his Australian birds for an equivalent set from the United States, thereby inaugurating an ongoing mechanism for exchange.[48] In some cases, the Australian director could not supply the specimens requested, but that seems not to have deterred the Smithsonian from sending frequent and larger contributions from its own large storehouse of specimens.[49] Despite or perhaps because of their scarcity, most Australian marsupials, for example, had acquired a familiarity that increased their value to Americans.[50]

The intermittent scientific exchange and displays were probably less directly influential in arousing public interest than were the growing imperial aspirations of Americans. In 1900, historian and publisher Herbert Howe Bancroft wrote a book entitled *The New Pacific* which, without embarrassment, portrayed the entire Pacific Basin as a region lying between two great 'English speaking and English thinking and acting peoples' who would inevitably shape its destiny, and he called for a greater connection between the United States and Australia.[51] The balance of authority and the nature of the proposed co-operation in economics, diplomacy and intellectual life remained ambiguous.

AUSTRALIAN-AMERICAN CONNECTIONS AND COLLABORATION

By the turn of the century, Australians had become more conscious of their similarities to and differences from the federated United States. Australian loyalty remained fundamentally with the British, but analogies to American experience became more common as Australia moved toward federation and Commonwealth status. Sometimes the comparison was intended as a spur. George Verdon, President of the Trustees at the National Museum of Victoria, observed in 1892, ' ... it seems appropriate to notice the absence of large private gifts and bequests in aid of most of the public institutions in Victoria. In America, colleges, museums, libraries and observatories are founded and maintained by the munificence of private citizens while in Victoria the gifts to this institution have been confined for the most part to books, a few pictures ... [with] none for building or endowment ... '[52] Verdon's comments, directed at local philanthropists in Melbourne, raised the important issue of sponsorship and identified one of the basic reasons for the frequent imbalance in exchange of specimens and publications. Australian museums were remarkably under-funded by comparison with American museums. Recognition of often similar goals and yet dissimilar means became fundamental to the relationship that developed after the turn of the century, shaped collaborative projects, and encouraged foundation spending.

Some international connections were forged in the 1880s, a decade when Australian museums grew in size and stature, when local initiatives overrode earlier British imperial incentives, and when numerous new American museums were established. Interchanges of books, honorary memberships, and specimens reflected an ever increasing internationalism in scientific efforts, stimulating contact across numerous national boundaries, including the great Pacific divide between the United States and Australia. Not all natural history exchanges were through museums, of course, and even within museums conducted by the Smithsonian, books recording the exchange of specimens list individuals and such government agencies as departments of mines and agriculture as well as museums.[53]

Sometimes, however, enthusiasm for exchange was muted by competition, particularly among Australians who observed that by the turn of the century the continent seemed 'overrun by the Collectors and Agents of Foreign Museums, dealers, and wealthy dilettanti' who acquired new and valuable specimens from unexplored regions. Some complained that wealthy Americans 'snap up every [rare zoological] specimen, & this has run up the market price'.[54] To these critics it appeared that unique and valuable aboriginal objects were being lost from Australia.[55] Little would be done, however, for several decades.

Crises in the 1890s, including depression and natural disasters of drought and flood, temporarily disrupted museum activities in Australia. Only government sponsorship insured their survival. By the twentieth century, urban developments and native-born leadership renewed efforts to establish autonomous institutions with distinctive programmes and fresh connections with other countries, including the United States. Historians have, for example, pointed out the symbolic significance of the American Great White Fleet that visited Australian port cities to an enthusiastic reception in 1908.[56] Just two years later an American, Walter Birley Griffin, won the competition to design Australia's new capital city, Canberra. In the Australian museums, staffs were rebuilt and even expanded. First Charles Dixson and then Charles Hedley, curators of the Australian Museum, travelled abroad to survey foreign museums.

Hedley's summary of the trip in 1911 highlighted the importance of paid staff, educational programmes and efficient methods for displaying and lighting specimens.[57] Both of their reports, though different in form and specific content, indicate that while the British Museum was still 'mother', the best practical museum models for Australia were in America. The image of two similar, independent countries nonetheless existed in tension with the reality that it was the British Association for the Advancement of Science that visited Australia in 1914, the British Royal Society that elected Australians to membership, and the British Museum whose collections continued to be consulted by Australian curators. Nationalism may also have been more immediately important than international connections to many Australian naturalists who sought to establish their authority on their own continent.

Australian natural scientists were regularly reminded that their continent remained a still under-explored resource and that they lacked the facilities to pursue important projects. In the 1920s they supported the new Council for Scientific and Industrial Research (CSIR) and sought to curtail the loss of potential research materials. When British and American scientists visited Australia, local museum staff were usually friendly but worried among themselves about the export of Australian objects and specimens. One early leader who advocated closer control over foreign scientists was W. Baldwin Spencer, McCoy's successor at the National Museum of Victoria. He had initially exchanged 'duplicate' Aboriginal material for Native American material with Franz Boas at the American Museum of Natural History in New York and with the Field Museum staff in Chicago.[58] As the years passed, however, Spencer increasingly advocated conservationist methods and, as director of the National Museum, told his colleagues in 1920, 'I would like to enter a strong protest against the sending to Europe of collections that can well be described here ... [local museums] are quite as competent to describe and take care of Australian type specimens as are our own colleagues in Europe'.[59] His concern intensified when, in the following year, the American Museum of New York decided to send the Whitney South Seas Expedition to acquire 'representative collections of Australian zoology and anthropology', despite Henry Fairfield Osborne's assurance that 'the standard research material belongs in the British and Australian museums'.[60] Shortly thereafter the British Museum wrote to the Commonwealth government asking permission to collect Australian fauna, to study them in London, and to retain type specimens. These external initiatives moved Australian museum curators and trustees to action.

In 1922 the Board of Governors of the Public Library, Museum, and Art Gallery of South Australia sent a resolution to museum directors in other states:

> The Board of Governors notes with some apprehension the frequent presence of oversea [sic] collectors with the object of obtaining examples of the fast disappearing Australian Fauna in some of its most characteristic and unique forms, and wishes to call the attention of kindred institutions to the need that effective restrictions should be placed on the operations of such collectors both as to such Zoological examples which they may consider should be fully protected and also a definite limit be placed on the number taken of other species which it is thought desirable to be placed under restrictions.
>
> It is suggested that the Director or a Zoologist of each Public Museum in the Australian States, together with representatives from the Universities in each case, should be appointed as consultative committees to advise the respective state governments as to what limitations should be imposed on such collectors.[61]

Positive responses to the letter from South Australia led other trustees and scientists to a meeting of museum administrators at the Pan-Pacific Science Congress held in Melbourne in 1923. Out of that session came a series of resolutions regarding permits to hunt, a list of animals for controlled export, the requirement of depositing types of new specimens and duplicates of rare specimens in Australian museums, and the need for co-operation (and elimination of overlap) among museum and university researchers in the several states. These resolutions were only recommendations, however, and in 1924 representatives from South Australia, New South Wales, Queensland and Tasmania met in Adelaide to discuss general issues of museum management and to consider possible legislative proposals. H.C. Richards, a professor from Queensland and a trustee of the museum there, recommended the formulation of 'a unified scheme for procedures governing the permission to collect, and to export, protected animals' that would be put into effect by the Commonwealth as well as state governments.[62] The expanded regulations increased federal and state oversight, primarily by requiring negotiation between outside researchers and their Australian counterparts. American expeditions would henceforth be under careful scrutiny.

Physical loss of materials was one problem, another was authority of a more subtle and complex sort regarding the research and educational functions of natural history museums. The extraordinary wealth and opportunities of American museums made them difficult to emulate and, moreover, many research staff members were sceptical about the tactic of gaining public support through education and popularization. Not everyone agreed with Hedley's positive assessment of American museum developments. Reporting on an internal debate at the Australian Museum that had gone public in 1926, the press concluded, 'The modernists have won a partial victory. But they have won it at a cost. Some complain that the institution has been "Americanized". Others are certain that it has not been Americanized enough ...'[63]

'Americanization' with its emphasis on education and public display (and an apparent de-emphasis on research) received an additional boost in the 1930s. The initiative came once more from across the Pacific. Andrew Carnegie, the wealthy steel magnate turned philanthropist, had in 1911 added to his original foundation funds explicitly for Canada, the United Kingdom, and British colonies, and had thus provided opportunities that were eventually extended to the Commonwealth countries of South Africa, Australia, and New Zealand. As in other Carnegie programmes, most of the funds went to educational research, library development, and adult education. Then, in 1930, the Carnegie United Kingdom Trust recommended $ 30,000 for the British Museums Association to conduct a general survey of the museums in the British dominions. Out of this came a series of volumes, including one by S.F. Markham and H.C. Richards entitled *A Report on the Museums and Galleries of Australia*.[64] The report to the Carnegie Foundation of New York provided a candid, critical and potentially embarrassing look at particular museums, as it identified declining physical facilities, unattractive displays and untrained staff. Indeed, the authors found a crisis situation, brought on 'with the force of a typhoon' by the Depression. Some scientific society museums had been swept out of existence while 'the peeling walls, the antiquated cases, and the pitifully small scientific staffs, all bear testimony to the ferocity of the gale' in even the large state museums.[65]

Markham, honorary secretary of the Museums Association of Great Britain, and Richards, from Queensland, recommended, among other things, a visiting grants scheme to permit Australian staff to visit museums abroad and thus learn up-to-date techniques. The report also highlighted the professional isolation of museum staff – there had never been a national association and few curators had even been to other Australian museums – and suggested that staff would benefit from greater communication and exchange among themselves. His interest apparently sparked by the report, Frederick Paul Keppel, President of the Carnegie Corporation from 1922 to 1941, visited Australia and New Zealand in

1935 and summarized his own opinions in an essay entitled, 'The Southern Dominions of the British Empire'. He, too, was struck by the lack of local revenue and responsibility in the centralized system of education and, while concluding that museums were better supported than libraries, nonetheless observed that museums 'do not work together and in a large degree lack professional direction'.[66] Comparing the Australian museums to those in the United States he observed that 'Private philanthropy has played a smaller part in the building up of institutions than with us', although a few people had been generous. Like Markham and Richards, he believed that it was important to bring the dominion museum staffs into greater contact with those in the United States and Britain and also closer to each other.

The Australians took advantage of the foundation's interest. H.C. Richards, a graduate of the University of Melbourne and professor of geology at the University of Queensland, co-ordinated a meeting held in Melbourne on 4 May 1936 for the 'general consideration of museum problems'. Unlike the meeting in 1923, which had stressed research issues and depended on trustee initiative, this second Melbourne meeting was attended by directors concerned about administrative issues. Thanks to careful planning and prior correspondence among museum directors, the latter meeting resulted in the formation of the Australian and New Zealand Art Galleries and Museums Association, which held its first official meeting in Auckland the following year.[67] Very shortly thereafter, the Association asked that a skilled preparator be sent to teach museum staff display techniques. In Richards's estimation, as in Hedley's, the Americans offered considerably more expertise in this area than the British. Frank Tose, Chief of Exhibits for the California Academy of Sciences, visited Adelaide and other major cities. He then conducted a class for preparators at the Australian Museum in Sydney. They constructed three habitat groups.[68] In 1937, development grants were given to all six state museums of natural history (as well as to several art galleries), the largest going to the National Museum of Victoria and the smallest to the Tasmanian Museum and Art Gallery.[69] As a result, the Australian Museum made several local natural history films, the National Museum of Victoria developed three dioramas, the South Australian Museum established a children's section, and the Queensland Museum purchased casts and models and also developed an educational service for primary and secondary schools.[70]

Over the next few years, a number of Australian museum staff members also visited the United States on Carnegie travel grants (see Appendix A.) These provided opportunities to learn from fellow administrators abroad, especially for staff like the beleaguered Ludwig Glaubert. During the war years he was the only paid scientist on the staff of the Western Australian Museum in Perth and apparently purchased books for the museum from his own small income. For those hoping to direct their colleagues towards the 'development of the museum along modern lines', the trip gave a unique opportunity to learn first-hand about 'gallery display, public lecture schemes, publicity, museum broadcasting, storage and general care, preservation and cataloging of collections, and most of all the public educational side of museum work'.[71] Most grantees were able to visit several American museums, not all of which had resources that dramatically outdistanced their own.

While the small grants had helped, Australian museums suffered during the Depression and the Second World War. Richards continued to be a principal force in the Association and spent a full year visiting American museums in 1943.[72] During the war years, however, the Carnegie Corporation 'gradually reduced its interest in small grants of these types', and after the war decided to spend more of its money in the United States and for 'men and institutions who have ideas along with force and skill to advance them'.[73] Post-Second World War published reports are far less explicit about the grants being given, although in 1947, after a visit to Australia and New Zealand, the director of the Carnegie Corporation made 'various grants' to institutions.[74] The lack of sustained support, however, meant that Carnegie's philanthropy had only a temporary effect on the programmes of individual museums, and it remains unclear how much the visits contributed to any long-term interaction between Australians and Americans.[75] What they did do was raise staff morale at Australian museums, provide incentives for community outreach, and establish a national organization along the lines of the American Museums Association that had been founded in 1906.

Joint scientific activities also became more common. Americans collaborated when they did research in Australia, but most often the joint enterprise had unequal participation and an unequal distribution of the results. The best example may be the Harvard Australian Expedition, which was actually a series of expeditions by the staff of the Museum of Corporative Zoology. Six staff members collected materials in their specialties over the course of 1931 and 1932. Herbert Lyman Clark, who had worked earlier to identify Australian starfish and sea urchins, visited again in 1929 and 1932 under grants from Carnegie and the Australian National Research Council. His field work, combined with loans and donations from obliging curators at each of the major Australian museums, resulted in the standard reference, *Echinoderms from Australia*.[76] When palaeontologist William C. Schevill explored northern Australia for the Museum in 1932, he invited Harold Fletcher of the Australian Museum to accompany him; but the latter was refused permission to go by his trustees and the Queensland Museum was unwilling (or perhaps unable) to participate officially. Nonetheless the Queensland Museum's curator, Heber A. Longman, generously showed the Americans to his best fossil sites and thus permitted some important specimens to go abroad. One result of the trip was that the first articulated skeleton of the gigantic fossil reptile *Kronosaurus* which – despite the poignant fact that Longman himself had first described it from a scrap of jawbone – went to the Museum of Comparative Zoology.[77] A British colleague of Longman sympathized: 'I wish you could find some benefactor to enable you to continue and work out these discoveries [on Cretaceous and Jurassic vertebrates] in the same way as the Americans do ... '[78] From the point of view of the MCZ's director, Thomas Barbour, the expeditions had been the most productive in 'establishing friendly relations with colleagues in Australia and in securing material for many lines of research' in over half a century.[79] At its conclusion he maintained that 'in several branches of the animal kingdom, at least, the [Australian] collection is more complete [at the MCZ] than may be found in any other institution in the world'.[80] Such Americans were insensitive to the ambiguous response of Australians whose potential research materials were being taken. Schevill, for example, commented to a colleague that 'Longman himself seems quite a good fellow but ... gives me the impression theat he will try to collar anything magnolious[sic] we may happen to get if he possible can. In that connection I found it extremely hard to get permission to take three kangaroos'.[81] Yet without the slightest tinge of conscience, Schevill had taken away the only full skeleton of the *Kronosaurus*.

State and federal laws, however, put museum staff in a position to control efforts of foreign philanthropists like New York's Richard Archbold, one of the last of the 'economic royalists ... enthusiastic about Museum safaris'. Archbold himself led expeditions to Papua New Guinea and also sent collector Gabriele Neuhauser to Australia to gather specimens for the American Museum of Natural History.[82] Although zoologist Ellis Troughton at the Australian Museum was frustrated by Archbold's exporting of mammals and wanted to limit Neuhauser's permission to collect, Longman helped her gain a permit. He made no demands that specimens be shared, but did request that any paratypical specimens of new species of mammals sent abroad be returned to Queensland. This the American museum agreed to do.[83] That practice, in fact, seems to have become the standard one. Australian museum staff supported foreign scientists by helping with customs arrangements for necessary supplies and the release of certain specimens, and in exchange they obtained the return of duplicate and paratypical specimens.[84]

Against a changing political and economic backdrop, Australian and American naturalists established bilateral and indeed multinational scientific connections that had mutual if uneven benefits to themselves and their institutions. American scientists, with their financial resources and intermittent enthusiasms, initiated activity that sometimes enhanced but may also have inhibited the work of Australians. Alexander Hrdlicka, a physical anthropologist from the Smithsonian, collected in the Northern Territory and probably spurred Baldwin Spencer to do his own outstanding work there.[85] In other instances the Americans came so well equipped that their presence may have inhibited and overwhelmed the initiatives of under-funded scientists like palaeontologist Herber Longman, who could not compete with his well-funded Harvard counterparts. The Carnegie grants, small and directed though they were, enabled contacts between American and Australian museum staff and, equally important, created an unprecedented national network within Australia. By 1945, bilateral connections between institutions in the United States and Australia emphasized personnel and methodology (by contrast with specimens and undifferentiated publications in the nineteenth century) and were extended through special attention by the National Science Foundation in the 1960s and later.[86] A stronger intra-Australian network put museum staff in a position to negotiate their international activity, even when confronted with colleagues from larger and more powerful institutions.

In the years following the Second World War, UNESCO established a division on museums, the activities of which were based on the assumption that museums were a particularly appropriate way to educate people about other cultures. The Pacific Science Congress of 1949 also recommended a committee on Museums in Pacific Research, perhaps in part an effort to balance the attention being paid to display and education and to focus on the scientific strengths of member countries. Formal international activities signalled the extent to which the isolation, and to some extent the autonomy, of museums had been supplanted by standardized procedures to encourage public participation and to protect cultural property against illicit transfer.[87] The co-operation among museums still faintly echoed but in many ways was far removed from the imperial system that had initiated Australian museums just a century earlier. The two sibling nations had matured and established their own quite independent relationship.

History of Science and Technology Program
University of Minnesota
U.S.A.

APPENDIX A

Carnegie Travel Grants to Australian Museum Staff

1936–7	N. B. Tindale	South Australian Museum, Adelaide
1938	Ludwig Glaubert	Western Australian Museum, Perth
1938	J. Roy Kinghorn	Australian Museum, Sydney
19??	A. R. Penfield	Museum of Science and Industry, Sydney
19??	E. O. G. Scott	
1939	H. M. Hale	South Australian Museum, Adelaide
1940	J. Kingsley	Australian Museum, Sydney
1940	P. C. R. Boswell	[Queen Victoria Museum], Launceston

Data derived from H.C. Richards, Presidential Reports (typescripts), National Library of Australia.

NOTES

* The author thanks R.W. Home, Marc Rothenberg, Linden Gilbank and others for their comments on this paper. Research on Australian museums was made possible by a Fulbright fellowship, and on American museums by a Smithsonian Institution Senior Fellowship.

1 Lucille Brockway traces imperial practice in *Science and Colonial Experience: The*

Role of the British Royal Botanical Gardens (New York, 1979); for geology see Robert A. Stafford, 'Geological Surveys, Mineral Discoveries, and British Expansion, 1835–1871', *Journal of Imperial and Commonwealth History*, 12 (1984), 5–32. On the models used to discuss colonial science and imperial science, see Roy MacLeod, 'On Visiting the "Moving Metropolis": Reflections on the Architecture of Imperial Science', in *Scientific Colonialism: A Cross-Cultural Comparison*, ed. Nathan Reingold and Marc Rothenberg (Washington, 1987), pp. 217–250. Centre-periphery studies contributed to comparative studies in language supplied by Edward Shils, 'Center and Periphery', in *Logic of Personal Knowledge* (Glencoe, Illinois, 1961), pp. 117–30.

2 Seymour Martin Lipset, 'The "Newness" of the New Nation', *The Comparative Approach to American History*, ed. C. Vann Woodward (New York, 1968); R. Else-Mitchell, 'American Influences on Australian Nationhood', *Journal of the Royal Australian Historical Society*, 62 (1976), 9–19; Norman Harper, ed., *Australia and the United States* (Melbourne, 1971); Ray Aitcheson, *Americans in Australia* (New York, 1972); Norman Bartlett, *Australia and America through 200 Years* (Sydney, 1976). Else-Mitchell points to the relationships among labor organizers in the United States and Australia and notes that Karl Marx, Mark Twain, and contemporary journalists all commented on parallels and similarities between the two places.

3 Anthony Trollope, *Australia and New Zealand*, 2 vols. (London, 1873), vol. 2, pp. 253–4. He added, as well, that 'to many leading colonists the idea of any divergence is altogether distasteful'.

4 Population statistics make the disparity quite evident. In 1861, Australia had 1,166,000 people while in 1860 the United States had 31,442,000 in 1901/1900, Australia had 3,774,000 and the USA 75,995,000; and in 1954/1950, Australia had 8,987,000 and the USA 150,697,000. Data from B.R. Mitchell, *International Historical Statistics* (Detroit, 1983), pp. 47, 50.

5 On colonial American science see Raymond P. Stearns, *Science in the British Colonies of North America* (Urbana, 1970). Early museums are discussed in Joel Orosz, 'Curators and Culture: An Interpretive History of the Museum Movement in America, 1773–1870' (PhD thesis, Case Western Reserve University, 1986) and Sally Gregory Kohlstedt, 'Collections and Cabinets: Natural History Museums on Campus, to 1860', *Isis*, 79 (1988), 405–26.

6 Phrase is from Linden Gilbank, 'The Acclimatization Society of Victoria', *Victorian Historical Journal*, 51 (1980), 255–70.

7 William Howitt, *The History of Discovery in Australia, Tasmania and New Zealand, from the Earliest Date to the Present Day*, 2 vols. (London, 1865). Missionaries had their own distinctive view of Australia, as illustrated in Daniel Wheeler, *Extracts from the letters and Journal of Daniel Wheeler, while engaged in a religious visit to the inhabitants of some of the islands of the Pacific Ocean. Van Diemen's Land, New South Wales ...* (Philadelphia, 1840).

8 Charles Coleman Sellers, *Mr. Peale's Museum: Charles Willson Peale and the First Popular Museum of Natural History and Art* (New York, 1980), p. 63. The list of *Donations to the Cabinets of the Essex County Natural History Society from January 1st to September 1st 1836*, p. 43, indicates a number of insects, reptiles, skins and other materials from Swan River, Western Australia, contributed by John B. Williams,

and some plants from Sidney [sic], New South Wales, from Charles F. Richardson. Britain's allocation of exclusive trading rights to British corporations and tensions before and during the War of 1812 made eastern Australian ports less common stopping places than those in New Zealand (to 1840), other South Sea Islands, and Perth. See Ernest S. Dodge, *New England and the South Seas* (Cambridge, Mass., 1965), pp. 182–3, and Werner Levi, *American-Australian Relations* (Minneapolis, 1947), esp. pp. 1–36; J.J. Eddy, *Britain and the Australian Colonies, 1818–1831* (Oxford, 1969), pp. 67–68. Fishing vessels in fact outnumbered American trading vessels sailing to Sydney, Hobart and Perth, according to Gordon Greenwood, *Early American-Australian Relations from the Arrival of the Spaniards in America to the Close of 1830* (Melbourne, 1944), p. 69.

[9] Vincent Pondo, Jr., *Ships, Seas and Scientists: United States Naval Explorations and Discovery in the Nineteenth Century* (Annapolis, 1974), p. 22. E. Daniel Potts and Annette Potts, *Young America and Australian Gold* (Brisbane, 1974), chap. 1, 'The Official Americans: The Consuls', pp. 6–24; also Norman Harper, 'A Historical Perspective', in *Pacific Orbit* (Melbourne, 1968), p. 177. Elena Grainger, *The Remarkable Reverend Clarke: The Life and Times of the Father of Australian Geology* (Melbourne, 1982), p. 92; Ann Mozley Moyal, *Scientists in Nineteenth Century Australia: A Documentary History* (Melbourne, 1976), pp. 90–91; T.G. Vallance, 'Pioneers and Healers: A Record of Australian Palaeontology in the Nineteenth Century', *Alcheringa*, 2 (1978), 246. The Boston Society was proud of its Australian birds acquired for comparison and display: Charles Sprague to S.F. Baird, 24 April 1856; RU 52, Box 10, Smithsonian Institution Archives (hereafter SIA).

[10] John Cassin to Baird, 21 August 1847; RU 7002, SIA and 'John Gould in America', *Contributions to the History of North American Natural History* (London, 1983), 51–8. Cassin claimed that the collection had all but five known species of Australian birds and that Gould pledged to secure those as well. On 24 April 1856, Charles Sprague reported to S.F. Baird that the Philadelphia Academy had again 'received from Victoria a splendid donation of Australian birds', Baird MSS, RU 52, Box 10, STA. The Academy of Natural Sciences also has Gould's personal copy of the *Catalogue of Specimens of Natural History and Miscellaneous Curiosities Deposited in the Australian Museum* (Sydney, 1837).

[11] A hand-drafted list of institutional correspondents, undated but from about 1862, lists several dozen European contacts but none in Australia; Agassiz MSS, Museum of Comparative Zoology, Cambridge, Massachusetts (hereafter MCZ).

[12] Royal Society of Tasmania, abstract of letter to the Colonial Secretary, 16 June 1860; Letter Book, Royal Society of Tasmania, Hobart (hereafter RST). The letter requested a grant of £ 100 a year for two years, plus expenses, for a pupil (possibly Albert Bickmore) of Prof. Agassiz from New York. Agassiz also proposed sending a Mr Garrett to Australia, New Guinea, and other islands in a letter of 30 May 1863 (Agassiz MSS, MCZ).

[13] See McCoy to Louis Agassiz, 25 January 1866, Letterbook II, and Alexander Agassiz to F. McCoy, 24 October 1866, Letterbox (U.S. Museums), Museum of Victoria, Melbourne (hereafter MVM).

[14] Alexander Agassiz to Newall & Co., 25 February 1865 and to Wilkinson & Bros., 25 February 1865; Agassiz MSS, MCZ.

[15] McCoy to Secretary of the Smithsonian Institution, 28 September 1860; Letterbook, MVM.
[16] Joseph Henry to F. McCoy, 6 February 1877, Letterbook (U.S. Museum), and F.B. Meek, then at the State Geological Office in Illinois, to McCoy, 23 May 1869, Letterbox, MVM; and abstract of a letter to the Secretary of the Smithsonian Institution, 8 October 1870, Letterbook, RST.
[17] Marsh to F. McCoy, 1 December 1871; Letterbox, MVM.
[18] Levi, op. cit. (n. 8), pp. 68–71, and Sherman F. Denton, *Incidents of A Collector's Rambles in Australia, New Zealand and New Guinea* (Boston, 1889).
[19] *Science*, 4 (1884), 497.
[20] According to his 'line a day' journal, Ward used Melbourne as a base from 25 April to 1 July while he collected regional specimens and tried to sell his exhibition. For the next three months, he moved north through Sydney, Brisbane, Rockingham and Port Darwin, also gathering Australian materials for his American clients. This journal is in the Henry A. Ward MSS, Special Collections, Rush Rhees Library, University of Rochester, Rochester, New York (hereafter RUR).
[21] A brief account by Ward is in the *Rochester Democrat and Chronicle*, 9 May 1882, WARD MSS, RUR. Ward was unable to sell an herbarium of McCoy, valued at $ 3,000, in the United States because buyers found it lacking in rarities and incomplete. See Ward to McCoy, 20 July 1882, 7 February 1885, and 8 December 1888; Ward MSS, RUR.
[22] The term 'fossil continent' was the title for an article from *Cornhill Magazine* and reprinted in *Popular Science Monthly*, 33 (1888), 682–93. The presumptuous title 'United States of the Pacific' – perhaps an awkward effort to express camaraderie – was used for an article on Australia in *Chautauqua Magazine*, 12 (1887), 17.
[23] According to a copy of a contract dated 6 January 1896, Ward was to be paid $ 250 plus travel and expenses for the work, which was to be done between Brisbane and Thursday Island in Torres Strait. Ward's enthusiastic proposal and a set of detailed letters from Ward to Agassiz, dated 15 August, 3 September, 15 October and 23 December 1986 document the work done by Ward and his assistant Mr. Potter, all in the Agassiz MSS, MCZ.
[24] For an analysis of Agassiz's own theory of coral reefs see Sir John Murray, 'Alexander Agassiz: His Life and Scientific Work', *Museum of Comparative Zoology Bulletin*, 54 (1911), 139–58, and Alfred Goldsborough Mayer, 'Alexander Agassiz', *Popular Science Monthly*, 77 (1910), 419–46.
[25] Georgiana King to Frederick McCoy, 5 April 1896; Letterbook, MVM. On the difficult life of this peripheral scientist, see 'Georgiana King: Amateur Geologist and Anthropologist, 1845–1932', *University of Sydney Archives: Record*, 9 (1982), 4–9. On the coral investigations see Roy Macleod, 'Imperial Reflections in the Southern Seas: The Funafuti Expeditions, 1896–1904', in *Nature in its Greatest Extent: Western Science in the Pacific*, ed. Roy Macleod and Philip F. Rehbock (Honolulu, 1988).
[26] Sybil Jack, 'Cultural Transmission: Science and Society to 1850' in *Australian Science in the Making*, ed. R.W. Home (Melbourne, 1988). On Australian museums see J.G. Marshall, 'Australian Museums: A Preliminary Bibliography to 1982', *Australian Historical Bibliography, Bulletin*, 8 (1983), and S.G. Kohlstedt, 'Historical

Records in Australian Museums of Natural History', ibid., 10 (1984), 61–82. F.G. Clarke, *The Land of Contrarieties: British Attitudes to the Australian Colonies, 1828–1855* (Melbourne, 1977).

[27] Quoted in Lionel Gilbert, 'Botanical Investigation of New South Wales, 1811–1890' (PhD thesis, University of New England, 1971), p. 538.

[28] T.L. Mitchell to Rev. George E. Turner, Secretary, Australian Museum, 14 May 1852; Correspondence to 1883, Series E, Australian Museum, Sydney (hereafter AMS). Mitchell donated 160 specimens collected in 'Tropical Australia' and identified by Dr Lindley, Sir William Hooker, Mr Bentham and Professor DeVries, thus indicating that the originals had been sent to England. See Michael Osborne on acclimatization in 'The Société Zoologique d'Acclimatation and the New French Empire: The Science and Political Economy of Economic Zoology During the Second Empire' (PhD dissertation, University of Wisconsin, 1987), pp. 364–70.

[29] Commenting on the problems of intellectual life, W. Stanley Jevons told his sister that in Australia his personal education would be 'checked and irretrievably deferred', and returned to England; in Australia he had gathered and published meteorological data but once back in England he turned his attention to political economy. Letter of 9 July 1958 in *Life and Journal of W. Stanley Jevons*, ed. Harriet Jevons (London, 1886), p. 103.

[30] Letters of introduction from R. Eglesfeld Griffith of Philadelphia to Mahlon Dickinson, Secretary of the Navy, and Com. D.S. Patterson, for W.S. Macleay, 10 May 1836, Macarthur Family Papers, Mitchell Library, Sydney (hereafter MLS). The Macleay Museum was, in 1890, transferred to the University of Sydney.

[31] See the discussion of Gerard Krefft's dismissal in Ronald Strahan, *Rare and Curious Specimens: An Illustrated History of the Australian Museum, 1827–1979* (Sydney, 1983), pp. 27–36. Earlier William Wall, for example, wrote to C. Blyth, Curator of the Asiatic Society Museum in Calcutta, that he lacked a suitable building and the means to make exchanges; see his letter, [?] March 1846, Correspondence Files (to 1883), and the Minute Book, 1836–1863, both in AMS. The Minute Book entry for 7 July 1846 indicated that the 'friends of the Institution' could not meet the expense of hiring a collector.

[32] Minute Book, 3 May 1856, AMS. His list reflected how little the Australians in Sydney knew of the museums being established in several major cities and, equally important, on college campuses.

[33] Copy of letter from George Bennett to the 1862 International Exhibition organizers in London, 6 June 1861, and Minute Book entries for 4 October 1856, 4 December 1956 and 2 June 1858, AMS. Macleay demonstrated a copy of the British Museum catalogue and the two curators, George French Angus and W.S. Wall, were asked to prepare catalogues of molluscs and birds respectively.

[34] The careers of these independent collectors suggest remarkable enterprise; for example, Oliver Haldane Stokes indicated having worked for the Royal College of Surgeons and the Brazilian Museum in a letter to the Australian Museum, Series E, 7 August 1871, AMS.

[35] Newall and Co. to DeVis, 16 January 1884; Incoming Letterbook, Queensland Museum, Brisbane (hereafter QMB). Apparently the only direct ships came from New England during the wool season.

[36] Sally Gregory Kohlstedt, 'Natural Heritage: Securing Australian Materials in 19th Century Museums', *Museums Australia* (July, 1984), 15–22.
[37] Potts and Potts, op. cit. (n. 9), p. 148.
[38] See the list, dated 28 September 1858, as part of McCoy's 'Catalogue and List of Specimens Sent or Received', Letterbox, MVM.
[39] Alexander Agassiz to Frederick McCoy, 24 October 1866, Letterbox and the record of exchanges in the annual reports (on Baird see 3 August 1880 and 2 February 1881), MVM.
[40] Roswell Ward, *Henry A. Ward*, p. 214, indicates that the collection was purchased by the Australian government, but the National Museum of Victoria's steering committee authorized only a small payment to Ward of £ 73 for specimens on 1 June 1882; MVM.
[41] Entry for July 1896, South Australian Exchange Book, 1888–1921, Royal Society of South Australia, Adelaide (hereafter RSSA). In exchange the Australian museum sent skins, boomerangs and other ethnological items.
[42] This was a relatively small number compared to the 622 listed for Germany and the 357 for Great Britain in the Smithsonian Institution's *Annual Report* (1874), p. 22.
[43] *A Catalogue of the Scientific Books in the Library of the Royal Society of New South Wales* (Sydney, 1889) indicates almost no books by American naturalists except those published by the Smithsonian.
[44] Agnew to the Secretary of the Smithsonian Institution, 22 March 1872; Letter Book, RST.
[45] Charles W. DeVis in Queensland Museum, *Report for the Board of Trustees for the Year 1885* (Queensland, 1886), p. 2.
[46] See E. Jewett to S.F. Baird, 22 March 1872 and 19 May 1873; Baird MSS, RU 52, Box 33, SIA. Jewett was a private collector who sent materials to Hobart Town, Tasmania, to Dr James Hector in New Zealand, and to Sidney [sic], New South Wales through the Smithsonian's exchange system. It was cheaper to send via Panama than overland via San Francisco.
[47] The first major effort to introduce European ideas about display was made by Gerard Krefft, 'The Improvements Effected in Modern Museums in Europe and Australia', Royal Society of New South Wales, *Transactions*, 2 (1868), 15–25.
[48] Discussion was initiated by the commissioner, but all subsequent contact was with the Australian Museum. See S.F. Baird to Charles Robinson, 28 August 1876, and the 'List of North American Birds Received from the Smithsonian Institution', 10 April 1877; Correspondence to 1883, AMS.
[49] A list of exchanges with the Australian Museum and a summary of letters indicating the Australian Museum's inability to send fishes from R. Etheridge, Jr., 1 August 1884, and an inability to send vertebrate fossils, are in records of Edward P. Ramsey, 26 November 1891; RU 189, Box 5, SIA.
[50] In 1901 W.N. Mincke offered a duckbill platypus for sale to Richard Rathbun at the Smithsonian Institution for \$ 250 to \$ 300, according to a letter of 8 January 1901; RU 189, Box 83, SIA.
[51] Bancroft included Canada as neighbor to the United States and New Zealand next to Australia on the two 'ends' of the basin, in his curiously projected frontispiece map of the Pacific. Bancroft, *The New Pacific* (New York, 1900), p. 223.

52 National Museum, *Annual Report for 1892* (Melbourne, 1893), p. 4.

53 See, for example, the United States National Museum, *Annual Report ... ending June 1895* (Washington, 1897), pp. 28 and 29, which indicates that the Smithsonian exchanged foraminifera with Edgar J. Bradley of the Happy Valley Waterworks, South Australia, and fossil invertebrates and plants with the department of mines and agriculture in New South Wales.

54 G.P. Whitney, quoting the Australian Museum's annual report for 1895, *Early History of Australian Zoology* (Sydney, 1970), pp. x and 9. See also Tasmanian H.H. Scott to W.B. Spencer, 29 May 1903; Letterbox (Australian Museum), MVM.

55 Thomas S. Dixson, *Australian Museum, Sydney, Lecture on Its Origins, Growth and Work* (Sydney, 1919) pointed out in frustration that the Bishop Museum in Honolulu had the first 'family scene' or habitat-styled group of Australian aborigines. Similarly the Gregory Mathews ornithological collection left Australia in 1918 because Australians were unable to raise a £ 10,000 purchase price; it went originally to the Tring Museum near London and later to the American Museum of Natural History. See R.T.M. Pescott, *Collections of a Century: The First Hundred Years of the National Museum of Victoria* (Melbourne, 1954), p. 117.

56 M. Ruth McGaw, 'Australia and the Great White Fleet, 1908', *Journal of the Royal Australian Historical Society*, 56 (1970), 121–31.

57 Charles Hedley, *Report on the Museum Administration in the United States*, Misc. Series 8 (Sydney, 1913); Hedley also visited European museums but did not include them in his published report, perhaps because Charles Anderson had already published his *Report on a Visit to Certain European Museums in 1911* (Sydney, 1912). Travel for self-education of museum administrators became increasingly common; see Sally Gregory Kohlstedt, 'International Exchange and National Style: A View of Natural History Museums in the United States, 1860–1900', in Reingold and Rothenberg, op. cit. (n. 1), pp. 167–90.

58 See Spencer's correspondence, including letter from Franz Boas dated 1 June 1903, in the National Museum Letterboxes, particularly those labelled American Museum and United States Museums, MVM.

59 Quoted in D.J. Mulvaney, 'Patron and Client: The Web of Intellectual Kinship in Australian Anthropology', Reingold and Rothenberg, op. cit. (n. 1), p. 69.

60 Henry Fairfield Osborne to Spencer, 1 April 1921; Spencer MSS, MVM. The British also continued to have Australian expeditions, see for example, George Herbert Wilkens, *Undiscovered Australia: Being an Account of an Expedition to Tropical Australia to Collect Specimens of the Rare Native Fauna for the British Museum, 1923–1925* (New York, 1929).

61 Invitational letter to representatives of museums and universities in Australia from the General Secretary of the Public Library, Museum, and Art Gallery of South Australia, 5 July 1923; in Australian Museum Correspondence Box, MVM.

62 H.H. Scott, typescript, 'A Short Account of the Informal Meeting of Museum Representatives' dated 28 August 1924 and found in box on Australian Museum Correspondence, MVM.

63 Strahan, op. cit. (n. 31), p. 68.

64 Stephen H. Stackpole, *Carnegie Corporation Commonwealth Program, 1911–1969*, (New York, 1963), pp. 11–13. For an interpretive study see Ellen

Condcliffe Lagemann, *The Politics of Knowledge: The Carnegie Corporation of New York, Philanthropy, and Public Policy* (Middletown, 1989). The report was published by the Carnegie Corporation in 1933, while the Museums Association published a *Directory of Museums and Art Galleries in Australia and New Zealand* in the following year.

65 Markham and Richards, *Report on the Museums*, p. 8.

66 Keppel, *Philanthropy and Learning with Other Papers* (New York, 1936), pp. 139–54, quotation on p. 141. Keppel apparently visited from January to June, 1935.

67 Carnegie Corporation of New York, *Report of the President for 1936* (New York, 1936), p. 32. Apparently H.C. Richards had special funds from Carnegie to provide travel costs for a special conference of the directors of museums; see South Australia Museum, Museum Committee Minutes, South Australia Public Library, Adelaide. A report by Markham and Richards on New Zealand, published the same year and often bound with the Australian report, concluded that New Zealand museums had a more impressive base of support and a tradition of more enterprising curators.

68 H.C. Richards to D.J. Mahoney, 2 August 1837, MVM. Copies of the 26-week itinerary of Tose are in a letter from D.J. Mahoney to F.P. Keppel, 1 October 1937, MVM. The staff from Brisbane, Adelaide, Melbourne and Hobart reported back on the excitement and long hours, as in F.A.A. Peake to D.J. Mahoney, 1 and 4 November 1937, MVM.

69 Stackpole, *Carnegie Corporation Commonwealth Program*, pp. 34–39, provides a list of the grants given by institution, date and amount.

70 H.C. Richards, Presidential Report to the Art Galleries and Museums Association (typescript), dated 13 January 1939, copy in National Library of Australia. The displays and films remained, but the educational programme at the Queensland Museum ended when the funds ran out; Patricia Mather, *A Time for a Museum: The History of the Queensland Museum, 1862–1986* (Brisbane, 1986), pp. 111–12.

71 This enthusiastic commentary was written by J.C. Kinghorn to J.R. Slevin, 26 June 1936; Slevin MSS, Department of Herpetology, California Academy of Sciences, San Francisco.

72 The National Library of Australia has typescript copies of speeches by Richards in 1939 and 1940 and a printed report for 1943.

73 Carnegie Corporation, *Report of the President and Treasurer for 1944*, p. 21, and *Report ... for 1946*, p. 15.

74 Carnegie Corporation of New York, *Report of the President and Treasurer for 1947*, p. 38.

75 This is in contrast to the Carnegie's importance in the United States, as discussed in Ellen Condcliffe Lageman, 'The Politics of Knowledge: The Carnegie Corporation and the Formulation of Public Policy', *History of Education Quarterly*, 27 (1987), 205–20.

76 Subtitled *An Account of Collections made in 1929 and 1932*, in MCZ *Memoirs*, 55 (1938). The introduction details Clark's gratitude and obligations to his illustrator wife, Frances L.S. Clark, and the numerous amateurs and curators in Australia who assisted him.

77 Mather, op. cit. (n. 70), pp. 140–141. The skeleton was not mounted, however, until 1959.

78 Arthur Smith Woodward to H.A. Longman, 27 January 1936, QMB.
79 Museum of Comparative Zoology, *Annual Report* (1931–1932), pp. 7–8.
80 Museum of Comparative Zoology, *Annual Report* (1932–1933), p. 8. Barbour added, 'I mean this literally ...
81 Schevill to Grover Allen, 20 February 1932, Allen MSS, MCZ. There are also journal notes (20 July 1931 to 23 January 1932) by Grover Allen at the MCZ Archives. A list of mammals collected by Schevill for the MCZ indicates 128 mammals, 217 birds, 800 reptiles and amphibians, a 'large number of insects and other arthropods' in 631 phials and pill boxes, and 2500 mounted insects.
82 Geoffrey Hellman, *Bones, Bankers and Beetles: The American Museum of Natural History* (Garden City, New York, 1968), p. 160.
83 Mather, op. cit. (n. 70), pp. 163–64.
84 J.R. Slevin to J.R. Kinghorn, 6 February 1929, C. Anderson to J.R. Slevin, 7 March 1939, and J.R. Kinghorn to J.R. Slevin, Slevin MSS, California Academy of Sciences, San Francisco.
85 Pescott, pp. 124–25. See, for some record of Spencer's work, *The Aboriginal Photographs of Baldwin Spencer* (Melbourne, 1982).
86 The annual NSF *Databooks* discuss bilateral cooperation in a general way and indicate that Australia has been one of relatively few targeted countries for such initiatives (including sponsorship of the conference that generated this volume of essays).
87 Edward P. Alexander, *Museums in Motion: An Introduction to the History and Function of Museums* (Nashville, 1979), pp. 245–7, and 'The Role of Museums in Pacific Research', Australian Galleries and Museums Association, *News Bulletin*, 3 (1949), 10–11. The author would like to thank Karl Kabelac of the Rochester University Archives and Alan Leviton of the California Academy of sciences for help in locating documentation for this paper. Much of the discussion on Australia reflects my on-going debt to colleagues and students at the University of Melbourne who helped me first work through the history of science in their country.

R.W. HOME

A WORLD-WIDE SCIENTIFIC NETWORK AND PATRONAGE SYSTEM:

Australian and Other 'Colonial' Fellows of the Royal Society of London

In recent years, several studies have focussed on the contribution of Britain's scientific institutions to the expansion of the British Empire in the nineteenth century. In particular, Lucile Brockway has stressed the importance of Kew Gardens in the establishment under Imperial control of important new cash crops such as cinchona, rubber and sisal, while Robert A. Stafford has emphasized the role of the Geological Survey of Great Britain and the Royal Geographical Society in promoting the discovery and exploitation of colonial natural resources.[1] As yet, however, no systematic analysis has been undertaken of the role of the nation's premier scientific society, the Royal Society of London, in the British imperial system.

That the Society played such a role cannot be doubted. On the one hand, there is abundant evidence that on matters scientific its advice was taken as authoritative by successive British governments, and that this advice was often directed to Imperial affairs.[2] On the other, as will be shown in this paper, election to membership of the Society was always held open to colonial scientists and came to be regarded as the ultimate accolade that could be bestowed on a scientist working in the colonies, just as it was for scientists working in Britain. The Society thus functioned as greater Britain's national scientific academy, the linch-pin of an empire-wide system of scientific patronage and reward that helped keep colonial science firmly bound to that of the metropolis. Even today, despite the break-up of the Empire, election to the fellowship remains open to scientists from the nations, now independent, of the Commonwealth, and significant numbers are elected, especially from Australia and Canada. (For example, there were four Australians in the 1988 intake.) In this way the Society has transformed itself from a 'national' into an international (or at least supranational) fellowship of scientists. Its policy in this regard has contributed to its maintaining its position as an institution of considerable international significance in a world in which, in most other respects, Britain's influence has in recent decades declined

R.W. Home and S.G. Kohlstedt (eds.), International Science and National Scientific Identity, 151–179.

dramatically. At the same time, the Society has in this way helped Britain retain a degree of cultural hegemony, at least so far as science is concerned, over its former colonial territories.

Our study thus raises in a particularly interesting way the question of nationalism *versus* internationalism in science. The scientific communities of the larger and more economically powerful nations of the Commonwealth have since the Second World War become self-sustaining in a way that they never were before, and it no longer makes sense, as it once did, to think of them as part of an extended British scientific community. On the other hand, with but few exceptions neither Britain's scientists nor those of the former colonial territories have wanted to sever the old links completely. The decision has been taken – positively and not merely through tardiness in changing the rule – that for Society purposes, residents of Commonwealth nations should *not* be treated as foreigners but should still be seen as part of the family. Only when a nation opts to leave the Commonwealth (or is forced out, as happened in the case of South Africa) do its residents become ineligible for election to the Society under the ordinary rules. The maintenance of this close link between scientists at the highest level has only occasionally been seen as incompatible with the newly achieved independent (or, better, non-dependent) scientific standing of the nations concerned. Likewise, it has apparently seldom been a matter of concern in Britain that a significant fraction of the ordinary membership of the national scientific academy, a principal source of scientific advice to the British government, should be made up of foreigners.

In this paper, I describe the pattern of elections of colonial Fellows to the Society in the heyday of empire, and chart the way in which the rules governing eligibility for membership have been adapted in more recent times as the old imperial links – or at least those of a formal kind – have gradually dissolved away. For present purposes, I focus particularly on the Society's Australian connections, this being the field I know best. However, I also present, in more abbreviated form, data concerning the election of Fellows from elsewhere in the Empire. I do so partly for purposes of comparison, but also because I believe the processes with which I am concerned are seen much better in an empire-wide than in a local context.

The Royal Society of London has, ever since its foundation in the seventeenth century, been an institution of more than merely local repute. Those living in foreign parts were welcomed as members from the outset and were elected in large numbers, the reading of communications from them formed part of the regular business of Society meetings, and many of these communications were subsequently published in the Society's *Philosophical Transactions*.

During the first half of the eighteenth century, foreigners were elected to the fellowship in such numbers that, in the 1740s, they represented almost 50% of the total.[6] Thereafter, controls were gradually introduced that made their election more difficult. In the process, the category of Foreign Member was formally defined for the first time. By 1787, the total number of such members was restricted to 100, and procedures were set in place whereby vacancies were filled once a year from among candidates who had been nominated during the preceding twelve months. Later, in 1823, the number of Foreign Members was further restricted to a total of 50, new members 'to be put in nomination at a meeting of the Council' instead of being selected at an ordinary meeting of the Society. Thereafter the rules governing the election of Foreign Members remained unchanged for many years.[7]

Of particular significance for our purposes here is the fact that these various restrictions only ever applied to foreigners. British subjects resident abroad continued to be eligible for membership in the usual way. Thus the earliest set of restrictions imposed, enacted in 1761, was stated to apply to 'persons residing in Foreign parts, not being subjects of the Crown of Great Britain',[8] while in the definitive rules adopted in 1787, the eligibility for ordinary membership of residents of Britain's overseas empire was made explicit, the new regulations applying only to persons 'who are neither natives nor inhabitants of his Majesty's dominions'.[9]

Not until after the Second World War did the definition of 'foreign' used by the Society in this context undergo further alteration. Hence, throughout the heyday of the second British empire, in the nineteenth and the first half of the twentieth century, any inhabitant of any part of the imperial territories, and any native Britisher living even beyond its borders, was eligible for election on the ordinary rather than the foreign list. A principal concern in this paper is the way in which this possibility was exploited in practice to maintain and strengthen the bonds of empire in the field of science.

It is well known that the ordinary criteria for election to the fellowship changed decisively during the course of the nineteenth century.[10] The effect of the change, which was incorporated in the revised statutes of 1847,[11] was gradually to transform the Society from a gentlemen's club for scientists and patrons of science into a national scientific academy, the membership of which was, with only a few exceptions, restricted to distinguished practitioners of one or other of the sciences. This was brought about by greatly restricting the number of new Fellows who could be elected each year – at first to only 15 per year – and by simultaneously ensuring that these were all elected on the same day, on the basis of a list selected by the Council of the Society from among those who had previously been recommended in writing by at least six Fellows. These new rules naturally affected the election of residents of the colonies to the fellowship, though not, it would appear, in such a way as to lessen the chances of success of the active scientists among them.

Among the new rules set out in the revised Statutes of 1847 was a requirement that, each year, a list of candidates together with the names of the Fellows who had proposed them should be printed and a copy sent to each Ordinary Fellow in advance of the election. The practice has continued to the present day. These lists, which are in effect printed summaries, undated, of the original certificates of candidature, also include details of each candidate's current position and a short summary, prepared by the principal sponsor, of the candidate's contribution to science that forms the basis of the nomination. They now constitute a very convenient starting point for the historian interested in membership patterns within the Society or in networks of influence within the larger British scientific community. The analysis that follows is based upon a survey of these printed lists covering the period from 1848, the first year of operation of the new system, to 1940, the latest year for which I was granted access.[12] For that span of almost a century, I have attempted to identify and extract details of the candidature of all candidates for election to the Ordinary List, successful or not, who were resident outside the British Isles at the time of their nomination. (Some had long since returned to Britain by the time they were elected.) I did not include in my survey the large number of candidates from Britain's colony in Ireland, though such an analysis could be undertaken quite straightforwardly once one decided whether or not to include candidates from Ulster in the count. I also extracted details in a much less systematic way for people resident in Britain at the time of their nomination who subsequently emigrated to one of the colonies, or who had built their scientific reputation overseas. However, definitional problems arise with candidates in this last category. For example, should Charles Darwin be included because of the importance to his career of his voyage on the *Beagle*? Or an explorer such as

James Clark Ross? In fact, they were not: I tried to include only those the bulk of whose scientific careers were spent abroad, or those who played an important part, themselves, in colonial life for a significant period. Without the overseas address as a marker, however, it is very easy to miss candidates in such categories. I suspect I have overlooked, in particular, many former members of the Indian Civil Service who on every objective criterion ought to have been counted.

No similarly convenient source as the printed lists of candidates exists for the period prior to 1848. For this, one faces the much more daunting task of working one's way instead through the bulky bound volumes of original certificates of candidature.[13] For this study, I confined my investigation of the pre-1848 elections to successful candidates who resided in Australia for a significant period, these generally being readily identifiable by an Australian historian such as myself in the chronological list of Fellows published in the Society's *Record*. It would be a much more laborious task to extract names of unsuccessful candidates from this period, though it was certainly the case that not all those who were nominated were elected.

The new arrangements set in place in 1847 had the incidental effect of more or less guaranteeing that, each year, some candidates would not be elected. At the same time, the introduction of the annual printed lists of candidates makes it a much easier task to identify who the unsuccessful ones were. In the first year of operation of the new rules, there were 22 candidates for the 15 places available. The vast expansion in science that occurred, both in Britain and throughout the Empire, during the Victorian era, together with the common practice of sponsors re-nominating many of the unsuccessful candidates from previous years, led to a steady increase in the number of candidates each year, and by the end of the century there were 90 candidates for the same number of places.[14] Not until the 1930s did the Society respond to the growing pressure upon it by increasing the number of places balloted for each year, raising this at first to 17 in 1930 and then to 20 in 1937. More recently the number of new Fellows to be elected annually has been further increased, to 25 in 1946, then to 32 in 1965 and to 40 in 1976,[15] but the competition for places remains intense.

Looking first, then, at the group of pre-1848 'Australian' Fellows, one finds – not surprisingly, in view of the fact that Australia was only settled by Europeans in 1788 and the number of settlers remained small – very few native-born Australians among them. In fact, there is only one,

Phillip Parker King, who was born on Norfolk Island in 1791 while his father was serving there as officer in charge, shortly after the arrival of the first European settlers in Australia.[16] Like his father, King joined the Royal Navy, where he served with distinction as a hydrographer and nautical surveyor, mapping extensive previously uncharted coasts in the north-west of Australia and in South America. He later settled in New South Wales where he was for several years resident commissioner of one of the colony's largest enterprises, the Australian Agricultural Company, during which time he often provided support and encouragement for other exploring expeditions visiting Australian waters. King was elected FRS in 1824 while resting in England between his two great surveying expeditions. On his certificate he was described, with much justice, as 'a Gentleman well versed in Geographical, hydrographical and astronomical knowledge who has recently completed a Nautical Survey of New South Wales'. Those supporting his candidature included Sir John Barrow from the Admiralty, Everard Home, Joseph Sabine and the politician Benjamin Hobhouse.

Almost all the other pre-1848 'Australian' Fellows were elected long before going to Australia and went there in an official capacity, as Governor (Bligh, FRS 1801; Brisbane, FRS 1810) or Colonial Secretary (McLeay, FRS 1809) of New South Wales, or Lieutenant-Governor (Franklin, FRS 1823; Eardley-Wilmot, FRS 1812; Denison, FRS 1836) or Colonial Secretary (Bicheno, FRS 1827) of Van Diemen's Land. Matthew Curling Friend (FRS 1820), a former naval officer, emigrated to Van Diemen's Land in 1832 and became port officer at Launceston where he gave occasional lectures on astronomy at the Mechanics' Institute. The only person elected FRS in this period while actually resident in Australia was J.H. Kay (FRS 1846), an officer on James Clark Ross' Antarctic expedition who remained behind in Hobart as director of the observatory established there in connection with the expedition's magnetic researches. His nomination was prepared in Hobart, presumably by Eardley-Wilmot, whose name heads the list of proposers, or Bicheno, these being the only signatories resident in the Australian colonies at the time; for obvious reasons it came entirely handwritten rather than on the usual printed form.[17] Robert Brown (FRS 1811) and John Gould (FRS 1843) were elected for contributions to Australian natural history based on extensive periods spent in Australia, but in both cases their election came only after they had returned to Britain and published their results there.

A very similar pattern would emerge, I am sure, with regard to

Britain's other colonial territories at this early period of British dominion. India would presumably yield a considerably larger number of British-born Fellows who at some time held positions in the colonial or East India Company administration, or who visited the country for the purpose of scientific exploration; but neither in India nor in the new 'settler' colonies (other than Québec, which had its own special problems) were there yet sufficiently large numbers of either settlers or expatriates temporarily in residence to generate a sizeable local scientific population from which a few leading members might expect to be elected FRS. The indigenous peoples being brought under British dominion would not, of course, at this early period of colonization have become sufficiently acculturated to Western modes of thought for any of their number to be considered as candidates.

Or so one might reasonably think. India, however, is always capable of springing surprises! On 27 May 1841 there was elected to the Society on the nomination of (among others) Barrow, Beaufort and Edward Sabine, one Ardaseer Cursetjee Esquire, 'Ship Builder of Bombay lately in England having undertaken the journey to this country at his own expense in order to prepare himself in the knowledge of the Steam Engine as applicable to Navigation and to acquaint himself with the arts and manufactures of Europe with the view of improving his own country and his countrymen, a Gentleman well versed in the theory and practice of Naval Architecture and devoted to scientific pursuits ... '.[18] Nevertheless, Cursetjee was very much an exception. Not until many years afterwards, in 1918, was another of his countrymen, the mathematician Srinivasa Ramanujan, elected.

It is, in fact, extremely unlikely that Cursetjee could have been elected under the rules adopted in 1847, for these strongly favoured the election of candidates who had an established record of scientific work, not just an enthusiasm for its practical application. There is, besides, a distinctly patronizing air about the description of Cursetjee's achievements, he being recorded as 'having introduced Lighting by Gas into Bombay where he perfected a small Gas establishment aided exclusively by Native workmen; having also at his own charge built a Vessel of sixty Tons to which he adapted a Steam Engine sent out from this country, and manufactured and fitted every other part of the Machinery and navigated the vessel entirely with native workmen and Engine men, chiefly instructed and trained by himself ... '. The phraseology employed and the nature of the achievements themselves suggest that very different expectations were brought to bear in considering Cursetjee's nomination than was usually the case. Under the new rules, where there was always an element present in the election of competition with the other nominees, there was much less chance of this happening.

During the first years of operation of the new rules, men were still occasionally elected whose achievements in the field of science were less than one might have expected. From the 'colonial' lists, examples are E.B. Eastwick (FRS 1851), professor of Hindi and oriental languages at the Indian Civil Service training college, Haileybury College, 'eminent as a philologist and linguist', and Sir John Bowring (FRS 1856), Governor of Hong Kong and 'eminent as a literary man, attached to science, and anxious to promote its progress'. As we shall see, a few other 'promoters' of colonial science were also elected later in the century. In time, however, the only non-scientists elected came to be those in the privileged class, defined initially to include 'a Prince of the Blood Royal, or a Peer of the United Kingdom, or one of her Majesty's Privy Council, [or] any Foreign Sovereign Prince, or the Son of a Sovereign Prince', whose right to proceed immediately to a vote for election following nomination was unaltered by the 1847 revisions.[19]

India and Australia were by far the largest colonial sources of candidates for election to the Society in the period under consideration, with India, the jewel of the second British empire, supplying a significantly greater number of successful candidates. Large numbers of Britishers built careers there in the colonial administration or elsewhere. During the second half of the nineteenth century, these included many scientists attracted to positions in the Indian government service at a time when there were still very few similar openings for professional scientists in Britain itself. Enough of these were elected to the Royal Society to constitute, eventually, a well-defined interest group that for several decades averaged close to one new Fellow elected each year. Probably no other group in British science did as well until the heady days of the Cavendish Laboratory under J.J. Thomson and his New Zealand-born successor, Ernest Rutherford.

Most of the Indian contingent appear to have been elected chiefly on the basis of their standing as senior members of government scientific agencies, rather than for the originality or intrinsic merit of their research. This, however, is by no means peculiar to the Indian group. While the system of election adopted by the Society in 1847 unquestionably had its intended effect of favouring, to the exclusion of almost all others, the candidature of those actively engaged in scientific work, it has never served to deny recognition to those whose major contribution to science has been on the administrative side. In the Indian case, the number of scientific agencies represented was considerable: botanic gardens in Calcutta and Ceylon, observatories in Travancore, Bombay and Madras, the forests service, the meteorological service, the Geological Survey of India, Calcutta's Imperial Museum, and the Bengal Medical Service. Most successful of all in having its staff elected was the Great Trigonometrical Survey of India, with no fewer than nine of its members being elected between 1858 and 1880 and others later, if not at quite the same rate.

Commencing with the election in 1889 of D.D. Cunningham, professor of physiology at Calcutta, one can trace the growing importance of India's recently established universities in the country's scientific life. Thereafter, a steady trickle of professors from Indian tertiary institutions appeared in the lists alongside, but never replacing, representatives of the government scientific services. All but one of the native Indians who appeared in the lists of candidates between then and 1940 had built their careers in the universities, and in fact they come to dominate the list of candidates from this source. (The exception was the extraordinary self-taught mathematician Ramanujan, elected after having been brought to England to work with G.H. Hardy.[20]) Senior positions in the government scientific agencies seem to have been reserved for British expatriates.

One effect of electing to the Royal Society so many senior officers of the scientifically orientated branches of the British administration in India was to enhance the scientific standing of those institutions. Another was to link these firmly, through the Society, with British scientific institutions. Likewise, as the universities developed, the election of some of their leading professors to the fellowship enhanced the British orientation and ethos of the entire Indian university system, establishing a British-based award as the ultimate criterion of academic status within the science-based departments. In this way, election to the Society provides a clear example of the use of science as an agent of European cultural imperialism, of the kind identified by Lewis Pyenson in relation to German overseas expansion in the early twentieth century.[21]

The pattern of Australian nominations is very different. The rules governing nomination demanded that certificates be 'signed by six or more Fellows, of whom three at least shall certify their recommendation from personal knowledge';[22] and, especially in the early days when there were very few Fellows resident in Australia (let alone in any single colony) the latter requirement could pose a major problem for 'home-grown' talent. In India, by contrast, the existence of a strong scientific bureaucracy, staffed by expatriates with good connections back in Britain and from an early date already containing a sprinkling of Fellows in its upper echelons, made it much easier to gather an appropriate group of signatories.

The Australian list thus has a much less organized appearance. Would-be candidates were forced to muster such local signatures as they could, even if those who signed knew next to nothing about their scientific work, and then to rely on the good offices of contacts in Britain to provide the backing from within their particular scientific specialties that quickly became an essential prerequisite to success. The candidature of the first Australian resident to be nominated under the new rules, the Tasmanian botanist R.C. Gunn, illustrates the problem perfectly. Gunn was elected at the first attempt, in 1854. His signatories included all three Fellows then living in Australia (Denison, King and Bicheno), J.H. Kay who had left Hobart just a few months earlier following his recall to England, Matthew Curling Friend, who had likewise returned to England shortly before, and the explorer Owen Stanley, who would have met Gunn when he visited Australia a few years previously.[23] (The name of J.D. Hooker, who visited Hobart as a member of Ross' Antarctic expedition and became friendly with Gunn, is at first glance a surprising omission from this list; he was a member of Council at the time and it seems already to have become established usage that Council members, well placed as they were to support favoured candidates in other ways, should not sign certificates.) Not one of those who signed 'from personal knowledge' was eminent in Gunn's own scientific field. His standing as a botanist was attested instead by those who signed 'from general knowledge', namely W.J. Hooker – with whom he had corresponded regularly for many years – Robert Brown, John Bennett and J.E. Gray.

The next Australian resident to be nominated, Captain Henry Ward, R.E., Master of the Sydney Mint, was less successful. With but one published paper listed in his application, and with a less distinguished set of sponsors than had been assembled on Gunn's behalf, he gave up after four attempts (1856, 1859–61). Several other Australian colonists subsequently had similarly pedestrian (and similarly unsuccessful) applications lodged on their behalf. Unfamiliarity – of Australians with the standard of the competition they were entering, often of their

Australian-based sponsors with the current international state of the fields in which expertise was claimed, usually of their British-based sponsors with anything but such of their work as had been published in Europe – seems to have been a major factor here. In India, many more scientists were employed, appointments tended to be for a limited number of years, and significant numbers of scientists moved between Britain and India each year. Scientists on each side thus kept reasonably well informed about the condition of science on the other. In Australia, by contrast, the numbers engaged were much smaller, appointments tended to be open-ended, and interchanges of personnel and the exchanges of information that accompanied them were correspondingly fewer.

In at least one case, the failure of a nomination can be clearly attributed to misunderstanding due to a combination of unfamiliarity with the rules and poor communication due to distance. W.P. Wilson, foundation professor of mathematics and natural philosophy at the University of Melbourne, was the prime mover behind the growth of the Melbourne Observatory in the 1850s and 1860s and, in particular, its acquisition of a major instrument, a 48-inch reflector, the Great Melbourne Telescope, one of the largest telescopes in the world at the time of its installation in the late 1860s. In his capacity as secretary of the Observatory's Board of Visitors, Wilson corresponded regularly with Edward Sabine, by then President of the Royal Society, during the protracted negotiations over the commitment to and manufacture of the telescope. Sabine was obviously impressed by Wilson's handling of the affair and in 1867, perhaps to mark the successful outcome of the telescope negotiations, arranged for his name to be put forward for election to the Society. Those who signed from personal knowledge included Wilson's former Cambridge tutors – he had been First Wrangler in 1847 – and Lord Rosse, whom he had known during the four years he spent as professor of mathematics at the new university college at Belfast in the early 1850s, before moving to Australia. In the absence of more detailed information, Wilson's scientific attainments could be stated only in the most general of terms, he being described as 'eminent in mathematical and physical science' and 'conspicuous for his efforts to promote in the Colony of Victoria the love of knowledge and to create the means of acquiring it'. Because the mails took so long to travel between England and Australia, Wilson's embarrassed response did not arrive in time to prevent his candidature going forward:

I certainly should consider it a great honour to belong to that body, and if I may interpret your letter as an expression of your opinion that I am entitled to it I gladly accept your offer. I have however always been of opinion that *original* scientific work of some kind was necessary, and my time has been so fully occupied in keeping myself up with the recent rapid advances in all branches of physical science, under the difficulties inseparable from residence in Australia that I have not been able to give sufficient exclusive attention to any one branch of science. I have always considered it my first duty to keep my teaching up to the level of the most recent discoveries in the various branches which fall to my share; and, as the whole mathematical, physical and astronomical teaching falls upon me singlehanded, I have found it no easy task. I enclose one of my recent examination papers, from which you may gather something as to what the University here is doing.[24]

Wilson was not elected that year; and in the light of his letter, his sponsors did not put his name forward again. How could they have nominated someone whose closest approach to a recent scientific publication was, by his own confession, the undergraduate examination papers he had set? Yet Wilson, through his unfamiliarity with Royal Society affairs and lack of close personal contacts in London to advise him, in fact had done his candidacy less than justice. To be sure, even in the most favourable of circumstances, he may not have been elected; but he had the President of the Society on his side and, contrary to his belief, by no means all those elected to the fellowship succeeded on the basis of their record as researchers. As we have seen, senior scientific administrators could also gain election. So, too, occasionally, could active promoters of science, as was proved by the election in 1864 of Sir Henry Barkly, the scientifically inclined Governor of Victoria who had, shortly before, taken up a similar post in Mauritius, and in 1870 of G.F. Verdon, a former Minister of Finance in the Victorian government who had been instrumental in finding funds for the building up of the Melbourne Observatory and for supporting science in the colony more generally.

The election of Ferdinand Mueller in 1861, at the second attempt, was an important stage in the Royal Society's links with Australia. Mueller, Victoria's Government Botanist and Director of the Melbourne Botanic Gardens, was the first person elected from Australia while holding a locally-funded scientific appointment, and for over a decade he remained the only Fellow working actively in the colonies as a scientist. He was thus, in effect, the Society's sole link during this period with Australian science and, in particular, with the expanding government scientific services.[25] Mueller had sought nomination from W.J. Hooker, director of Kew Gardens, with whom he had long had a regular correspondence, and

Hooker complied willingly enough, assembling, as other signatories, several other botanists including his son J.D. Hooker, William Harvey, Robert Wight, J.J. Bennett and John Lindley. Only Harvey had actually met Mueller; the others who signed 'from personal knowledge' evidently felt able to do so on the basis of their correspondence with him.

At the time of Mueller's election and for some years afterwards, any candidacy from Australia (or New Zealand) had to be managed in London from beginning to end. Thus the New Zealanders James Hector (FRS 1866) and Julius Haast (FRS 1867) were sponsored by a strong contingent of British-based naturalists and geologists including Murchison, Ramsay, J.D. Hooker, Huxley, Lyell and Jukes, without a single colonial signatory. Likewise, Wilson's nomination in 1867 was handled entirely in London, as was the unsuccessful candidacy in 1869 of the foundation professor of the Melbourne medical school, George Britton Halford. Beginning, however, with the election of the director of the Melbourne Observatory, R.L.J. Ellery, in 1873, some nominations of Australians began to be initiated in Australia. Once as many Australian signatures as possible had been collected, the certificates were sent to England. More signatures were added there before the documents were deposited at the Royal Society. Kay and Mueller signed Ellery's certificate, then sent it to England where such luminaries as Sabine, Balfour Stewart, Huggins, Humphrey Lloyd, J.C. Adams and W.J.M. Rankine subscribed. Mueller was the successful initiator of Frederick McCoy's successful candidacy in 1880; Ellery and Verdon signed for the unsuccessful candidacy of F.B. Miller of the Melbourne Mint in the same year; Verdon, Mueller and Ellery all sponsored the similarly unsuccessful candidacy of Sir Samuel Wilson of 'Ercildoune' and, a year later, that of John Day, past president of the Medical Society of Victoria, for whom various medical discoveries were claimed.

In all, Mueller appeared as a signatory for ten candidates from Australia and another two from New Zealand. Ellery was almost as active, signing eleven certificates including one from New Zealand. However, their nominees had a fairly low success rate, only five of Mueller's and four of Ellery's being elected. (There is, of course, considerable overlap between the two lists). Those whose nominations were managed in England did considerably better, scoring six successes out of nine candidates resident in Australia at the time of their nomination (and five out five among those resident in New Zealand) during the period in which Mueller was a Fellow. The difference is striking and suggests that Mueller and Ellery, both of whom constructed their scientific careers more or less from the outset in Australia and neither of whom ever revisited Europe thereafter, had difficulty in judging the criteria for election. Mueller certainly had his own doubts on this score, and when he sent a nomination for his friend and collaborator James Ruddall to J.D. Hooker, he explicitly asked Hooker not to put it forward unless he was sure of the support of the Council of the Society.[26] Moreover, some of those whom Mueller and Ellery sponsored do appear to have been of insufficient scientific standing. In addition, however, even their stronger nominations had the disadvantage of not being shepherded through the election by active supporters 'on the ground' in London. As argued earlier, unfamiliarity with work being done on the other side of the world doubtless also sometimes played a role. Mueller felt that the candidacy of the naturalist and Catholic priest, J.E. Tenison Woods, had suffered particularly on this count: '[I] can't see why 30 years labours in various branches of science with extensive fieldwork count for nothing', he complained.[27]

The pattern of Australian elections to the Royal Society during this period suggests that Australian scientific workers enjoyed marginal status, at best, in the imperial scientific system at this time. A few leading figures secured election to the Society, but even the major government scientific agencies for the most part lacked the firm institutional connections with London that their (usually much larger) Indian counterparts enjoyed. Essential to a successful election campaign, it appears, were well-established links with leading British figures in one's field, either through personal acquaintance (Cockle, Verdon) or chiefly established through correspondence (Gunn, Mueller, Ellery, W.B. Clarke, McCoy). Those candidates whose scientific links were chiefly within Australia (Day, Tenison Woods, P.H. MacGillivray, Ralph Tate) were uniformly unsuccessful, even though almost all had built up considerable local scientific reputations before their names were submitted to London.

As in the Indian case, the growing strength of Australian university science is reflected in the Australian list in the last two decades of the century. Indeed, the change is much more pronounced in the Australian case, with the universities coming to dominate the list so completely that in the sixty years from 1880 to 1940, only four of the twenty-five Fellows elected while holding Australian appointments came from outside the university system. Almost all had received most of their scientific training in Britain: the majority were British-born and had been educated there, while most of those who were brought up in Australia had at least undertaken postgraduate study 'at home'. The personal connections thus established with British colleagues show up clearly on the nomination certificates. To be sure, later candidates usually had among their sponsors some of their Australian colleagues who had been elected earlier; but in every case except that of the Sydney biologist W.A. Haswell (FRS 1897) there is a heavy preponderance of British-based support among the signatories on the certificates. Even Haswell's list included several British luminaries such as T.H. Huxley and E.R. Lankester.

The number of university scientists elected to the Royal Society in the years around 1900 served to lock the Australian universities very firmly indeed into the imperial scientific network. The universities were far smaller institutions than they are today, with only a few hundred students each and a handful of professors and lecturers to teach them. Yet no fewer than five professors or lecturers at the University of Melbourne (W.B. Spencer, J.W. Gregory, C.J. Martin, J.H. Michell and D.O. Masson) were elected FRS between 1900 and 1903, with their colleague T.R. Lyle following a few years later. The University of Sydney boasted three Fellows (Archibald Liversidge, W.A. Haswell and T. Edgeworth David) at this period with a fourth, J.T. Wilson, to follow soon afterwards,[28] while the University of Adelaide had one, E.C. Stirling, with W.H. Bragg to follow in 1907. And these were the only universities in Australia at the time, apart from the tiny University of Tasmania founded only a few years earlier!

The Australian universities had always been staffed in the main by British professors and, as W.P. Wilson's comments, quoted earlier, indicate, the scientists among them had always seen election to the Royal Society of London as the ultimate scientific accolade. However, few among the first generation of professors saw this as an honour to which they could realistically aspire. McCoy was the only one of their number to be elected, and he succeeded not on the basis of his university work but on account of his extra-mural contributions as director of the National Museum of Victoria and palaeontologist to the colony's Geological Survey. The election of so many Australian professors in the 1890s and 1900s made it clear, however, that election now was an attainable goal for university men, and success quickly became the hallmark, within the Australian university system, of scientific standing. Election came to be seen almost as the norm for those who were active researchers. Failure became deeply wounding and something of an embarrassment to one's sponsors, as can be seen, for example, in the correspondence of H.S. Carslaw, professor of mathematics at the University of Sydney. Carslaw, author of some important and justly esteemed work on Fourier transforms and their application to the theory of conduction of heat, had every reason to expect, on the basis of this work, to be elected. When he was not, the matter became a source of temporary difficulty between him and his friends who were sponsoring him. 'By the way', Carslaw eventually wrote to W.H. Young some months after his fourth successive failure at the election, 'you have never referred to the Royal Society. Forsyth quite expected that I w(oul)d be elected last year. ... And to tell the truth I was rather severely disappointed'.[29]

Inevitably, as the pressure on places increased inexorably, failures became more frequent. Success demanded not only an ever stronger record of research, but also good timing and careful marshalling of forces; and even then, election was no longer assured even after several years of candidacy. As early as the 1890s, places had come to be allocated notionally between the different scientific disciplines. New candidates tended to find themselves joining something of a queue of fellow specialists, and only the most outstanding (for example, Bragg in 1907) managed to leap-frog those ahead of them. It is on this basis that we find J.J. Thomson consoling Threlfall, who was a close friend, after his failure to get up in the 1898 elections, that 'there is no slight attached to not getting in the first time, as it is now very rarely indeed that a physicist does so; in some of the other departments of science it is a little more frequent as there is not such a press'.[30] So unusual did first-time election become, in fact, that in 1917 the Society changed its rules concerning renewal of candidature. Until then, renewal had required one of the proposers to request it annually in writing, whereas from that date,

> Any Candidate whose name shall have been printed in a previous list of Candidates, but who shall not have been elected, shall, unless his proposer or seconder withdraw his name before the 31st December immediately preceding an election, be a Candidate at such election. ... Provided always that the same certificate shall not be valid for more than five successive elections.[31]

A very clear picture of what came to be involved, both formally and informally, in organizing a nomination is given in a 1923 letter from L.N.G. Filon, FRS, to his friend Colonel L.R. Thomas, recently appointed Registrar of the University of Tasmania, in response to a feeler from Thomas about the possibility of nominating the University's longstanding mathematics professor, Alexander McAulay, a noted contributor to the early history of vector calculus:

putting up a candidate for the Royal is a ticklish business and is hardly worth trying unless there is a really good chance. I have had, for the first time, to put up a really good candidate this year, and I ought not to be putting up a second one, otherwise I shall prejudice the chances of both. Besides, one is statutorily debarred from putting up (as proposer or even seconder) a candidate whom one does not know *personally*. Therefore it is *vital*, if McAulay is to become a candidate, that we should know *at least three* Fellows (& those of some standing) who know him *personally*. There is, it is true, a rule that either of the two Secretaries shall have power to sign a certificate on behalf of a Fellow not resident in the U.K., on receiving a written request from him. So that if there are Fellows in your part of the world who know McAulay personally, the fact of their being far away could be got over. But, for any chance of success, there should be more than 3 Fellows who know the candidate personally & *at least* 6 signatures are required altogether. The first thing to do is to find a proposer. You say something of Major MacMahon. Does he know McAulay personally? If so, will you let me know and I will approach him. *He* might do the proposing & his word carries great weight. I know MacMahon and can speak to him. But it is useless to do so unless his knowledge of your friend is *personal*...

What is McAulay's age? This is important. Also are the papers 4 to 9 in your list accepted by the R.S. for publication? Has he done any research work before? I doubt whether papers 1 2 3, which appear to be the only ones published, would be likely, by themselves, to carry election.[32]

McAulay *had* done other research and so Thomas pursued the matter with various Australian Fellows. D.O. Masson from Melbourne was, by a fortunate chance, in England during part of 1924 and was able to approach senior mathematicians there on McAulay's behalf. By October that year, he was able to send Thomas from London a list of the signatures he had obtained:

It is now, I think, sufficiently representative and I am sending the certificate in, without further delay, with a personal letter to the Secretaries. It is impossible, of course to foretell the result; but, in any case, I hope McAulay will be pleased to know that he has the friendly support of every mathematician that has been approached. I may say that Jeans and Eddington, being members of Council, are barred; and of course they will have the chief say in this case.[33]

Those who had signed were J.H. Michell (proposer), T.R. Lyle (seconder), Masson, C.A. MacMahon and Joseph Larmor from personal knowledge and, from general knowledge, E.T. Whittaker, A.E.H. Love, H. Lamb and H.F. Baker. Masson had done well and assembled a list of signatories that carried considerable weight. It did not, however, carry enough to get McAulay elected in five elections prior to his death in 1930.

A question of considerable interest is whether, and if so to what extent, special allowance was made in the Society's elections for the special conditions under which candidates from the colonies were working. Certainly in the case of McAulay's candidature, there is no suggestion of any such thing; and in any case it is scarcely possible to divine what electors – or, more precisely, members of the discipline-based Sectional Committees[34] and of the Council of the Society, for it was they who drew up the list of preferred candidates each year – had in mind in judging the merits of any particular candidate. Nevertheless, there are indications that the exigencies of colonial life were occasionally taken into account.

First, one may point again to the election, even after 1847, of a few men from the colonial service who were distinguished for the promotion rather than the practice of science. The elections of Browning, Barkly and Verdon certainly come into this category; and similar considerations probably applied in the case of Sir William Jervois (FRS 1888), Governor of New Zealand, even though he was described on his certificate as 'distinguished as a military engineer', for it was also noted there that he had 'promoted the progress of Science in various ways'. And then there was Sir John Kirk, Her Majesty's Agent and Consul-General in Zanzibar, elected in 1887 at the first attempt, whose certificate noted how, after serving as naturalist on Livingstone's famous expedition, he had stayed on in Zanzibar for almost twenty years and had assisted many other exploring expeditions during that time. Election on such a basis under the ordinary rules seems to have been reserved for candidates from the colonies. Within Britain itself, distinguished promoters of science were catered for in a different way, under the rule governing elections within the privileged class. Following the revision of this rule in 1902, it could be applied equally to colonial promoters of science and therefore these appear no longer in the lists of ordinary candidates.

Again, it is fairly common for a nomination of a candidate from a colonial university to mention success in building up a school as one of the achievements worthy of note. W.P. Wilson's efforts to promote 'the love of knowledge' in Victoria were cited as one of his major contributions to science. Baldwin Spencer was said to be 'distinguished as an original investigator in zoology and comparative anatomy, and as a teacher and organiser'. Liversidge's 19 published papers were cited, but also his 'good work done in various ways for advancement of natural knowledge in Australia'. B.D. Steele (FRS 1919), in addition to earlier valuable work on the migration of ions in solution, 'has done splendid work in creating and organising the chemical department at Brisbane, but also the University of Queensland'. 'He has rendered signal service to the Empire in this respect', the report continued. T.H. Laby (FRS 1931), in addition to his own researches, 'has promoted research on Experimental Physics in Australia, and many researches have been published by students working in his Laboratory'. Otto Maass (FRS 1940), professor of physical chemistry at McGill University, in addition to publishing papers of his own, 'has trained many excellent graduate students, and has a reputation as the leading teacher of physical chemistry in Canada'.

In most such cases, the list of publications is also a respectable one and may have secured election in any event. What is noteworthy is that the additional statements about success in teaching are chiefly confined to applications from the Empire and seldom if ever appear on those of British-based candidates. Being a successful teacher of science in a more-or-less remote corner of the Empire seems to have been seen by at least some sponsors as more noteworthy than being one in Liverpool or Cork. One hesitates, however, to draw too definite a conclusion from evidence of this kind, since in many cases the person who actually drew up the statement – who, as we learn from the McAulay example, would not necessarily have been the person who appeared on the certificate as proposer – would have been a scientist from the colonies. The prevalence of statements about teaching performance may to some extent merely reflect a systematic misperception by colonial scientists as to what was required. What counted in the election, after all, was not what was said on the certificates but whether such statements carried weight when the Council of the Society came to draw up its list of recommended candidates each year.

In this regard, the candidature of B.D. Steele offers some evidence that, on at least some occasions, such statements may have helped. Steele was first nominated in 1909 and then again, always unsuccessfully, each year from 1911 to 1915. His certificate referred each time to his researches in physical chemistry alone. His candidacy lapsed in 1916 but was renewed the following year with a new certificate, on the basis of which he was eventually elected in 1919, Queensland's first FRS since James Cockle over half a century earlier. It was only on this second certificate that the statement appeared about his teaching and administration in Brisbane, and it could have been this that made the difference. However, he had also in the interim spent time in Britain helping to supervise the work of the war-time munitions factories, and the contacts he developed during this period may have been what swung the balance in his direction.

On only a couple of occasions did a certificate explicitly invoke (or appear to invoke) the difficulty of doing science in the colonies. In the case of the Indian physicist C.V. Raman (FRS 1924), we find the same patronizing attitude being aired towards things Indian as was noted earlier in respect of the nomination of Ardaseer Cursetjee. Raman, we are told with a hint of faint surprise, 'although trained entirely in India, has made considerable additions to our knowledge of sound and light...'. Some years later, those sponsoring the nomination of the New Zealand botanist J.E. Holloway (FRS 1937) wrote that 'the quality of his work ... is excellent and in view of the fact that his researches were carried out in very difficult circumstances the quantity is remarkable'. Those circumstances included, however, not only the fact that he was working in New Zealand but also that at the time he did most of the research he was a clerk in Holy Orders, actively engaged in parish work!

Quite apart from the question of whether one should give extra credit for work done well, if it was done in the colonies, at least one prominent Fellow is on record as holding the view that, all other things being equal, the Society ought occasionally to give preference to colonial scientists, simply because they were colonial. When D.O. Masson was first proposed, in 1901, William Ramsay endorsed his candidature in the following terms:

> Of all the chemical candidates I should rank him first, except perhaps, Smithells, who might be put level. Masson is professor of chemistry at Melbourne University, and has acted as consultant in all chemical questions to the Government of Victoria. He has not published much, but what he has published is good. He had a paper in Philosophical Transactions last year on ionic velocities which is a very pretty piece of work. *I think it is well to elect a colonial now and then*, and Masson ranks first I should say in our Australian colonies.[35]

It is of course possible that Ramsay was entirely alone in holding such a view. One cannot help suspecting, though, that his opinion was shared by others and that it played a part in the sudden surge of elections of Fellows from Australia, once the wave of talented new professors appointed as a result of the expansion of science teaching in the universities there had had time to build up reasonable programmes of research and thereby to establish their plausibility as candidates. The same years saw similar surges in elections from Canada and, to a lesser extent, South Africa, with eleven new Fellows and four respectively being elected between 1890 and 1910.[36]

There certainly was a more widely held concern within the Society at this period to strengthen links with the colonies, for a decade earlier the possibility had been canvassed – perhaps partly in response to the many unsuccessful nominations from the colonies, especially Australia, in the years before that – of creating a new, subsidiary category of Corresponding Members, to include foreign and colonial scientists who had not yet been elected Foreign Members or Fellows respectively.[37] Huxley was a strong supporter of the proposal, which was not, however, adopted:

> We shall look very foolish if, after all the talk, nothing is done to bring us into closer relation with the Colonial and American confreres (*sic*) ... I have not the slightest doubt that the Corresponding Membership would be highly appreciated both in America and in the Colonies and would afford deserved recognition to comparatively young men not only there but in this continent.[38]

In the long run, the rush of new elections from the colonies to the fellowship itself was surely a better solution. Thereafter, as the colonies concerned gradually established their political independence, they were able to maintain a significant representation in the fellowship of the Society, thus enabling the Society to claim, despite the continuing heavy preponderance of British-based Fellows, to be an institution of genuinely imperial scope.

The years after the Second World War saw rapid changes in the imperial structure as the old, settler-based Dominions – Canada, Australia, New Zealand, South Africa – more and more went their own way and the other former colonial territories one by one claimed their independence. These changes forced the Royal Society to change the rules governing its membership; but, as stated already, these changes did not by any means parallel those taking place in the political realm.

Traditionally, as we have seen, eligibility for election to the Society on the ordinary list had been defined, *de facto*, by the definition of eligibility for Foreign Membership, and has thus included all those who were 'natives or inhabitants of his Majesty's dominions'. Shortly after the Second World War, several important changes occurred in the notion of what constituted 'his Majesty's dominions'. Already, under the Statute of Westminster of 1932 that was ratified one by one by the old dominions during the course of the following decade or so, those dominions had been declared independent of Great Britain. Now both Ireland and Burma declared themselves foreign countries and thus definitely no longer subject to His Majesty. On the other hand, India and Pakistan, while also achieving independence, did so within the framework of what began at this time to be called the Commonwealth – even though India, by declaring itself a republic, no longer acknowledged the occupant of the British throne as head of state. (Pakistan later, however, in 1972, decided to leave, at which stage the status of its citizens changed; this has recently changed again with Pakistan's readmission to the group.)

In changing its rules in 1948 to accommodate these changes, the Society made the first, crucial decision, namely to maintain a broad basis of ordinary membership so far as the territories of the old empire were concerned. By relying on the notion of 'British subject', which under the British Nationality Act of 1948 included citizens of India, and by defining those eligible for Foreign Membership as those 'who are neither British subjects by birth nor British subjects by naturalization or marriage', the Society was able to maintain the eligibility of Indians for election to the ordinary fellowship while avoiding the problem that India was no longer part of His Majesty's dominions.[39] This, however, had the effect, perhaps unforeseen, of excluding citizens of the Irish Republic, even though they continued in Britain to enjoy almost all other rights of British citizens including the right to vote in British elections. The rule was therefore changed again in 1951, those to be considered for Ordinary Membership now being required to be 'a British subject or citizen of Eire'.[40]

Both of these new formulations, by relying on notions of citizenship rather than domicile, had the further effect of excluding, until such time as they took out British citizenship, the many eminent refugee scientists who had settled in Britain before, during and after the war. At least one commentator has suggested that this was a deliberate even if undeclared policy, designed to prevent the exclusion of worthy British candidates in the face of competition from the refugees.[41] Mere residency of a Commonwealth country did not again become sufficient to establish eligibility until 1979, and then only after heated debate within the Society. By then, however, the issue was no longer the status of the refugee scientists – if this had ever been an issue – but that of Americans working in Britain.

It was noted earlier that the number of places in the Royal Society to be filled each year has been raised on several occasions since the Second World War. Most recently, it was increased in 1976 from 32 to 40. At that time, there was general agreement within the Society that the very great increase that had occurred in the number of scientists working in Britain, combined with the continued eligibility of scientists from the Commonwealth, had undesirably restricted the chances of deserving British candidates being elected within a reasonable space of years. This generated a debate that must mark the end-point of our discussion. One school of thought within the Society wanted to solve the problem by transferring Commonwealth scientists to the Foreign List and thus effectively reducing the press for places on the Ordinary List. Another school, which had a resounding victory when the matter was put to the vote, preferred to reduce the pressure by increasing the number of places while keeping these open to scientists from Commonwealth nations. The argument that political independence in the former colonial territories should be accompanied by a severance of the ties that bound the science of those territories to London was thus firmly rejected.

Some years before this, during a Conference of Commonwealth Scientists sponsored by the Royal Society in 1967, the same issue was aired from a very different perspective when Australia's most eminent scientist, Frank Macfarlane Burnet, Nobel laureate in physiology or medicine in 1960, argued the case for national independence with officers of the Royal Society. Burnet approached the question from the point of view of the presidency of Australia's premier scientific institution, the Australian Academy of Science, a body modelled directly on the Royal Society and including all but two of the Fellows of the Society then resident in Australia among its membership. Burnet felt that, by continuing to elect significant numbers of Australians to the fellowship (but significantly fewer than the Australian Academy needed to elect each year in order to maintain itself) and by making every Australian scientist eligible for election, the Royal Society was in fact putting the Australian Academy in an invidious position. Since the membership of the Australian Academy was never confined to Australian Fellows of the Royal Society, it inevitably divided into those who were and those who, while eligible, were not (or were not yet) Fellows of the British body. In these circumstances, membership of the Australian Academy could not fail to be seen as second-best, a mere staging post on the way to election to the more exclusive London organization. Burnet became convinced that this arrangement was not in the long-term interest of the Australian Academy, and suggested that the Royal Society be asked to accept no further Australian nominations for election as Fellow. However, the proposal attracted little support either within the Academy or in Britain, either then or later.[42]

Yet Burnet was surely right in his reading of the situation. Unless the Royal Society changed its rules, the Australian Academy of Science and the equivalent bodies in the other countries of the Commonwealth were bound to continue playing second fiddle to London until such time as their local scientific communities became sufficiently strong, both numerically and qualitatively, to enable them to be at least as selective as the Royal Society – in fact, to reject some candidates whom the Royal Society accepted. Given the relative sizes and strengths of the scientific populations concerned, such a thing was likely to be some way off. In the meantime, Burnet's solution, with the decision to exclude being taken in London, was the only way in which the situation could be changed. Unlike the situation with, say, imperial honours, which are awarded on the recommendation of the government of the relevant country and can therefore be stopped (as has happened in Australia) by that government refraining from making new recommendations, election to the Society was, and is, a matter for the Society and the individual scientist alone. Nationalistically inclined individuals might decline to be nominated – and some have certainly done so – but there has never been a mechanism whereby any individual or group could decline on behalf of an entire national scientific community.

Whether Burnet was also right in regarding such an arrangement as something to be regretted is another issue. Would it matter if Australia's premier scientific institution were thus locked for the foreseeable future into a subservient position vis-à-vis its British equivalent? Is it merely a matter of national pride, or is there something else at stake? Perhaps it could even be argued to be in Australia's interest to see such an arrangement maintained, in that, by providing a standard of excellence and objectivity beyond Australia's shores at which the nation's scientists could aim, it would discourage an unduly inward-looking view of what counted as good work.

I can remember hearing similar arguments offered in my youth to justify the use in those days of the (British) Privy Council, rather than the nation's own High Court, as Australia's ultimate source of legal authority. Yet in the legal case, it would be difficult today to find a defender of that system. Similar reservations ought to be entertained, I believe, with regard to science, whether in the Australian case or in that of any other nation boasting a mature and self-supporting scientific system. Continuing to be locked into the British scientific hierarchy would result in British views being imposed not only as to quality but also as to which sciences and, within them, which topics of investigation, were most worthy of reward. And British views on such matters are neither infallible nor necessarily in accord with local experience. More generally, one questions why, in a world where science has become universal, it should be British views (rather than, say, Russian or Japanese or American views) that should prevail. Why, indeed, should not local views prevail?

Interestingly, in very recent times signs have begun to appear, at least in Australia – I do not know whether it has also been happening in the other nations of the old empire – that the traditional power relationship that Burnet sought to sever is breaking down of its own accord. During the Australian Academy of Science's early years, any Fellow of the Royal Society who chanced to take up an on-going appointment in Australia was virtually assured of being elected FAA at the first available opportunity. A comparison of the fellowship lists shows that on no fewer than nine occasions between 1954, the year of formation of the Academy, and 1976, an incoming FRS was elected FAA within eighteen months,[43] and on only one occasion, that of the physicist J.C. Ward (FRS 1965), who moved to Australia in 1966, was an incoming FRS not promptly elected to the Australian Academy. By the same token, only once during these same years was an Australian working in Australia elected to the Royal Society, who had not already been elected to the Australian Academy of Science. The exception was the director of the Snowy Mountains Authority, the engineer Sir William Hudson (FRS 1964), whose exclusion from the Australian Academy symbolized that body's persistent refusal to follow the Royal Society's example and treat pre-eminence in engineering practice as grounds for election to its fellowship – a policy that was eventually to lead to the formation of a new body, the Australian Academy of Technological Science (now the Australian Academy of Technological Sciences and Engineering) that would honour excellence in this field.

If the refusal to fall into line over the election of engineers was the first outward sign, in this context, of Australian scientific independence, Ward's failure to be elected FAA had more profound implications. Occurring at precisely the time of Burnet's onslaught on the Royal Society's procedures, it amounted to the Australian body denying automatic acceptance of the Royal Society's authority in judging the relative merits of candidates, even in fields of 'pure' science where the parallelism of interest was beyond dispute. It amounted, in other words, to a declaration of total independence of judgement in respect of scientific achievement. Admittedly, for over a decade Ward's remained the only case of its kind. More recently, however, there have been two other cases of Fellows of the Royal Society moving to positions in Australia but not being elected to the Australian Academy. Even more striking, since the late 1970s there have been several instances of Australian scientists, working in Australia, being elected to the Royal Society but not, or not so soon, to the Australian body. Furthermore, it is clear that these were not merely cases of oversight by the Australians, to be rectified as soon as they were revealed by the Royal Society's electing those concerned to its fellowship, since some have still not been elected to the Australian Academy, several years after being elected FRS.

To be sure, so long as the Royal Society maintains its policy of regarding scientists from the nations of the Commonwealth as eligible for election to its ordinary fellowship, one can expect that most of those elected will already have been elected to their own national scientific academy. What is more, this will insure that the traditional view, that election to the Royal Society is the higher honour, will continue to be widely held. Nevertheless, it does appear that, at least so far as Australia is concerned, in a technical sense the earlier pattern of subservience has been broken. No longer do Australians invariably look to London as the ultimate arbiter of scientific standing and authority. The painter has, if only recently, been cut.

Department of History and Philosophy of Science
University of Melbourne
Australia

NOTES

1 Lucile H. Brockway, *Science and Colonial Expansion: The Role of the British Royal Botanic Gardens* (New York, 1979); Robert A. Stafford, 'Geological Surveys, Mineral Discoveries, and British Expansion, 1835–71', *Journal of Imperial and Commonwealth History*, 12(1984), 5–32; idem, 'The Long Arm of London: Sir Roderick Murchison and Imperial Science in Australia', pp. 69–101 in R.W. Home, ed., *Australian Science in the Making* (Sydney, 1988); idem, *The Empire of Science: Sir Roderick Murchison, Scientific Exploration, and Victorian Imperialism* (Cambridge, 1989).

2 Marie Boas Hall, *All Scientists Now: The Royal Society in the Nineteenth Century* (Cambridge, 1984), especially pp. 166–74, 199–215.

3 R.P. Stearns, 'Colonial Fellows of the Royal Society of London, 1661–1788', *Notes and Records of the Royal Society of London*, 8(1951), 178–246.

4 Henry Lyons, *The Royal Society, 1660–1940: A History of its Administration under its Charters* (Cambridge, 1944), p. 53.

[5] *The Record of the Royal Society of London*, 4th ed. (London, 1940), p. 95.
[6] Lyons, op. cit. (n. 4), pp. 126, 343–4.
[7] *Record*, op. cit. (n. 5), p. 100.
[8] Ibid., p. 95. Foreigners resident in Britain thus continued to be eligible for ordinary membership, without restriction.
[9] Ibid., p. 96.
[10] Lyons, op. cit. (n. 4), chapters 7 and 8; Hall, op. cit. (n. 2), *passim*.
[11] *Record*, pp. 301–3.
[12] Royal Society, *Certificates of Candidature*, 1848–1940–.
[13] Royal Society, 'Certificates'.
[14] *Record*, p. 75.
[15] *Record*, p. 60; Royal Society, *Year Book*, 1940.
[16] Biographical information about King and other Australian nominees is drawn chiefly from the *Australian Dictionary of Biography*.
[17] Royal Society, 'Certificates 1840–1860', f.173. On Kay and the Hobart magnetic observatory, see A. Savours and A. McConnell, 'The History of the Rossbank Observatory, Tasmania', *Annals of Science*, 39(1982), 527–64. Friend, whose signature might also have been expected to appear on Kay's nomination, for some reason did not sign it.
[18] Royal Society, 'Certificates', IX.57.
[19] The rules governing elections within this privileged class were subsequently reconsidered. In 1865 the category of foreign princes embraced by them was somewhat extended, but then in 1874 the class was restricted to Princes of the Blood Royal and Privy Councillors. In 1902, the provision covering Privy Councillors was repealed and a New Statute passed to the effect that:

> In cases in which the Council is of opinion that, in the interests of the advancement of Natural Knowledge, it is desirable that persons be elected Fellows of the Society otherwise than (in the ordinary way), they may recommend to the Society for election persons, who, in their opinion, either have rendered conspicuous service to the cause of science, or are such that their election would be of signal benefit to the Society. Provided that not more than two persons shall be so recommended in any one calendar year, and if two persons be elected in any one year there shall be no election in the following year.

Under this Statute the South African Prime Minister, Jan Smuts, became a Fellow in 1930, as did the Australian statesman S.M. Bruce in 1944 and the Australian Prime Minister, R.G. Menzies, in 1965.
[20] *Dictionary of Scientific Biography*, XI, 267–9.
[21] Lewis Pyenson, *Cultural Imperialism and Exact Sciences: German Expansion Overseas, 1900–1930* (New York, 1985).
[22] *Record*, p. 301.
[23] Stanley died, at Sydney, in March 1850. However, the signature that appears on Gunn's certificate seems genuine. One can only conclude that it must have taken a very long time to collect the other supporting signatures in this case.
[24] Wilson to Sabine, 24 March 1866; Royal Society, Sabine correspondence, no. 1810.

[25] The mathematician James Cockle (FRS 1865 on the nomination of his mathematical friends in Cambridge), Chief Justice of Queensland since 1861, had few links with science outside that colony and chiefly social ones even within it. J.H. Kay, though now in a potentially influential position as secretary of Victoria's Executive Council, did not play an active role in that colony's scientific life. Gunn was isolated in Tasmania and increasingly involved in his non-scientific administrative responsibilities.

[26] A.M. Lucas, 'Ferdinand von Mueller, Protégé turned Patron', pp. 133–152 in Home, op. cit. (n. 1). Ruddall's nomination did not proceed.

[27] Ibid.

[28] Another Sydney professor, Richard Threlfall, was nominated in 1898 but resigned unexpectedly and returned to England before his election in the following year (R.W. Home, 'First Physicist of Australia: Richard Threlfall at the University of Sydney, 1886–1898', *Historical Records of Australian Science*, 6(3) (1986), 333–57).

[29] Carslaw to Young, 15 August 1915; University of Liverpool Archives, Young papers, D. 140/9/7.

[30] Thomson to Threlfall, 7 May 1898; Archives Department, Birmingham Reference Library, MS 347A/234 (quoted in Home, op. cit. [n. 28], p. 348).

[31] Royal Society, *Year Book*, 1918, pp. 59, 177. The period of validity of a certificate was extended in 1963 to seven years, but an interval of three successive elections thereafter then prescribed before a new certificate could be lodged, which certificate would be valid for only three elections before a further three-year intermission was required.

[32] L.N.G. Filon to L.R. Thomas, 1 December 1923; University of Tasmania Archives, UT 40/373.

[33] Masson to Thomas, 15 October 1924; University of Tasmania Archives, loc. cit.

[34] Established in 1896 with the responsibility, *inter alia*, of advising Council as to the merits of candidates from the different fields of science.

[35] Ramsay to W.M. Hicks, 17 March 1901; Ramsay papers, University College London (my emphasis); quoted by L.W. Weickhardt, *Masson of Melbourne* (Melbourne, 1989), p. 57.

[36] Though the number of South African elections seems small, it effectively trebled the number of South African Fellows. Similarly, there had been only four Canadian Fellows elected between 1848 and 1890, so that there, too, the increase was, proportionately, very large.

[37] Royal Society, *Council Minutes*, 15 March 1888.

[38] Quoted by Hall, op. cit. (n. 2), p. 123.

[39] Royal Society, *Year Book*, 1949, p. 84.

[40] Ibid., 1952, p. 84.

[41] L. Badash, *Kapitza, Rutherford, and the Kremlin* (New Haven, 1985), p. 36.

[42] Frank Fenner, 'Frank Macfarlane Burnet, 1899–1985', *Historical Records of Australian Science*, 7(1) (1987), 39–77; p. 59. A.L.G. Rees, personal communication.

[43] Two cases differ slightly from the rest. Otto Frankel (FRS 1953) arrived in Australia from New Zealand in 1951 with his election to the Royal Society already in train. This came too late, however, for him to be made a party to the negotiations that led to the formation of the Australian Academy and thus to be named automatically

along with the other FRSs resident in Australia (except A.G.M. Michell, who declined) as one of the Academy's Foundation Fellows. He was elected to the Academy in the following year. D.P. Craig was not an FRS when he returned to Australia in 1967 after many years in England, but he became one within a few months, in the Society's 1968 election. He arrived in Australia too late, however, to be considered by the Academy for its own 1968 election. He was elected FAA a year later.

One other case perhaps deserves comment, that of E.B. Verney (FRS 1936), who took up a personal chair at the University of Melbourne in 1961 following his retirement from the chair of pharmacology at Cambridge, but who was never elected FAA. Verney's Melbourne appointment was for a three-year term only, and it seems to have been generally anticipated that he would return to Cambridge, as he did, at the end of that time.

C. STEWART GILLMOR

IONOSPHERIC AND RADIO PHYSICS IN AUSTRALIAN SCIENCE SINCE THE EARLY DAYS

Ionospheric physics is a part of the international field of geophysics. Its history can be traced to the late nineteenth century, to the efforts of geomagneticians and to the first radio experiments. In 1902, in the very early days of radio, a high altitude conducting layer was suggested as a mechanism to allow for long distance propagation of radio signals travelling far around the curvature of the Earth. Theoretical efforts to explain this phenomenon attracted only moderate attention until about 1920. At this time, the development of radio vacuum tubes, piezo-electric crystal circuits, the amazing long-distance radio signalling feats of amateur 'ham' operators and the emergence of radio broadcasting turned world attention to questions of skipping, swinging and strays (long-distance radio propagation over 'dead' zones, fading of signals, and atmospheric interference).[1] Ionospheric physics was effectively established in 1925, when E.V. Appleton and M.A.F. Barnett in England and M.A. Tuve and G. Breit in the United States demonstrated that medium-frequency radio waves could be reflected off an ionized region or regions lying 100 kilometres or more above the Earth's surface. The ionosphere and its much larger outer region, the magnetosphere, extend from about 60 kilometres to thousands of kilometres above the Earth's surface – that is, from the lowest altitude where enough ions and electrons exist in the atmosphere to affect significantly the propagation of radio waves, up to the magnetopause, the boundary where the Earth's magnetic field ceases to control the dynamics of the Earth's atmosphere.

Commerce, civil governments and the military have constantly and from the beginning kept interest in, and supported radio and ionospheric physics. Geophysics, radio and radar astronomy and plasma physics have all used ionospheric physics as a tool to study the Earth's atmosphere and geospace. Indeed, radio study of the ionosphere has contributed to the birth of several famous offspring, including radar and radio astronomy. Pulse, FM and Doppler radar techniques were first developed to study the ionospheric layers. The first successful radio astronomy work, by Karl

R.W. Home and S.G. Kohlstedt (eds.), International Science and National Scientific Identity, 181–204.

Jansky, came in a survey of noise sources hampering transatlantic ionospherically propagated communications. Other ionosphere workers during the 1930s discovered and studied radio propagation during meteor showers, and radio noise during bursts from the Sun. Of course, many radar workers during the Second World War came from other backgrounds, but except for the Dutch, the early postwar radio astronomers tended to be British and Australians who had some study of the ionosphere in their background. Other ionosphere scientists and engineers developed forward scatter communications and over-the-horizon backscatter radar following World War II.[2] The International Union of Radio Science (URSI) was the first scientific body to recognize radio astronomy as a separate discipline, to formally urge support for rockets and satellites in scientific research, and to move to establish the International Geophysical Year, 1957–58 (IGY).

As in other countries, early radio development in Australia came partly from amateurs. It came also, however, from the seizing of research opportunities by teachers in Australia's institutions of higher education and by the development of Australia's telecommunications systems.[3] An Australian group about 1905 originally planned two high-power wireless stations near Sydney and Fremantle. Competing groups formed Amalgamated Wireless (Australasia) Ltd. (known as AWA) in 1913. Sir Ernest Fisk of AWA received wireless messages from Britain on 22 September 1918 sent from the Marconi station at Carnarvon, Wales, on a frequency of about 21 kHz (wavelength of 14,300 metres). Experiments continued on these long wavelengths, and in 1921 Britain agreed to co-operate with Australia's proposals for direct services. The so-called 'short wave' experimenting of the amateurs on wavelengths shorter than 200 metres brought forth the startling possibility of low power, higher frequency transmissions over very long distances. Fisk conducted 90 metre wavelength tests in February 1924, and on 8 April 1927 the 'Beam' system was operating to England on frequencies around 25 MHz (wavelengths of about 12 metres). Fisk was a great supporter of radio as well as of his company, AWA. The Post Master General's (PMG) Department wanted fully to control Australian radio. An expert consultant summoned in 1922 to help sort out the PMG's wireless problems stayed on permanently: Harry Percy Brown (later Sir Harry, and known as 'Poo-Bah' Brown) headed the PMG telecommunications from 1923 until 1939 and was an influential supporter of and financial contributor to radio research. (Brown had several different titles during his tenure.) One other government radio researcher should be mentioned: S.H. (Sid) Witt. Witt was a self-taught engineer and pioneer in government telecommunications who formed what would become the PMG 'Research Laboratories' in Melbourne in 1923 and headed it until 1945. Just as AWA was formed out of competing companies, so the British imperial cable systems combined in 1929 to form Cable and Wireless Ltd., to manage British Empire cable and radio telegraph and beam systems.

This radio activity attracted the interest of academics at the universities of Melbourne and Sydney. J.P.V. Madsen was the first professor of electrical engineering in an Australian university. He had studied physics and engineering at Sydney, then apprenticed under W.H. Bragg at Adelaide and later returned to Sydney to spend the rest of his career there. In Sydney, Madsen organized and splendidly led others rather than pursuing research himself. Madsen's counterpart T.H. Laby, likewise a former Sydney undergraduate and then the new professor of physics at Melbourne, had been one of the first Australian postgraduate students to study at the Cavendish Laboratory in Cambridge with J.J. Thomson. When he returned to Melbourne in 1915, Laby brought the Physics (or Natural Philosophy) Department into X-ray and then radio wave work. Laby was close to Ernest Rutherford and sent his best students, such as Harrie S.W. Massey, to study at the Cavendish. Lord Rutherford, the native New Zealander, maintained close ties with the antipodal lands and invited other students also to England. Thus he suggested that George H. Munro come to England from Auckland University. This had some weight as Munro's teacher, P.W. Burbidge, had been a student of Rutherford as well as of Laby.[4] As things would work out, Sydney University did more of its radio research from within Electrical Engineering. Melbourne University did radio research from within its Physics Department. Over the long haul, Madsen at Sydney would be more successful than Laby in generating a research enterprise and the radio and ionosphere work at Sydney would become the greater.[5]

In 1926 Laby obtained a grant of £ 500 from a local radio station for studies of radio field strengths. In that year, the Council for Scientific and Industrial Research was formed to undertake or support scientific research.[6] The organization had been aborning for over a decade and now its Chairman and central figure would be the well-known Sydney consulting engineer George Julius. The CSIR (as it was called then) would engage mostly in agricultural research but there would be one notable exception: radio research! Madsen approached his friend Julius and asked for support. The international radio science organization (URSI) was suggesting the formation of a national Australian liaison committee. The British had acted similarly several years previously in forming a Radio Research Board. What if the Australians could also form such a Board, under the CSIR, to guide radio research and administer financial support? Madsen pointed out that Professor Laby, AWA's Fisk, the PMG's Brown, the commercial radio stations in Sydney and Melbourne, the military, and the Wireless Institute of Australia all thought such a Board would be a good thing. The Board was approved and held its first meeting in June 1927. Madsen was chairman and included on the five-man board were Brown and Laby. Almost immediately afterwards, Madsen headed for England and the British Radio Research Board to seek advice and plan co-operative ventures. In 1928 Brown arranged for the PMG to give a grant of £ 8,000 to the Australian Radio Research Board (RRB). In fact, until 1943 the PMG funded about 70% of the total annual budget of the RRB.[7] Until the end of the Second World War, the university grants from the RRB went only to the Melbourne and Sydney groups.

Long distance radio was booming, the ionosphere was being probed and studied in America, Britain, Germany, France, Japan and elsewhere, and an Australian Radio Research Board had been formed.[8] What were the Australian radio workers now going to do? Radio was exploding as a communications medium and as a social force. Broadcasting journals appeared, radio companies were formed; even French perfumes were sold in bottles made to resemble vacuum tubes.[9] But customers, both commercial and the public, experienced interesting and frequently bothersome radio phenomena.[10] Radio static (called 'strays', or 'X's') of natural origin was mixed with man-made interference. Radio signals were subject to fading (also called 'swinging' or 'deviating') both of intensity and of fading rate. On some radio frequencies, stations could be heard reliably during the daytime out to distances of 80 kilometres or so but not beyond. At night, they could be heard over long distances as well as short, but with frequent severe fading at all distances. On other, higher frequencies, the patterns seemed reversed, with some high frequency circuits at long distance seeming better at night, others seeming better during the daytime. Some frequencies seemed to have concentric 'skip' or 'dead' zones wherein no signals were received in a radius of 800 kilometres or so from the transmitter. Static or 'strays' seemed to occur during rainy seasons and localized lightning storms were known to correlate with static. If an ionized upper layer or layers were responsible for assisting the radio waves to propagate, of what did this layer consist – ions? electrons? both? How high were these layers and how thick, and how did they vary over time and over different parts of the earth? Would radio waves 'bounce' off the upper ionized layer(s) or region or would they gradually be bent in passing through? Did they in fact come *down* from the layer or were they propagated in an oblique horizontal direction or reflected from low-lying mists or fog? Zeeman splitting was known to occur with electromagnetic waves in magnetic fields. Would not the same frequency splitting occur with radio waves passing through the earth's magnetic field? What kinds of interference were there in addition to known man-made noise and 'strays' – did any come from beyond the Earth?

There were instrumental problems as well as ones of propagation. Antenna development was (and is) a complicated mix of empirical and mathematical-theoretical work. Antennas became much, much more efficient than the old Marconi types of 1902 but the 'Beam' system still required antenna arrays covering many acres of ground and in some cases the arrays were over a mile long. Tuned radio frequency and Super-regenerative receivers were being replaced with Superheterodyne receivers. As operating frequencies went higher, atmospheric noise became less a design problem than noise generated within the receiver itself. Tubes were needed to operate at higher frequencies with lower noise factors (for receiving) and higher power output (for transmitting). New modulation methods were devised: amplitude modulation, frequency modulation, pulse modulation, double and single side band modulation. Australians would be among the world's most inventive in several areas of technique. David F. Martyn would partially achieve the technique of Doppler radar. Geoffrey Builder, as much as anyone in the British Empire, was responsible for developing pulse studies of the ionosphere.[11] W.N. Christiansen, B.Y. Mills, J.L. Pawsey, L.L. McCready and R.N. Bracewell produced fundamental work in the theory and design of antennas and in imaging theory.[12] Some of this work went on under auspices of the RRB and of its successor after the Second World War, the Radiophysics Laboratory in Sydney. Other work was sponsored by the AWA corporation.

Australian radio and ionosphere workers built upon work done elsewhere, especially in England. They developed and strengthened imperial communication links and links within Australia. They also exploited resources and their spatial location to study ionospheric, geomagnetic and astronomical subjects involved with radio. As examples, certain phenomena appear different due to the fact that Australia is in a different hemisphere from Europe. Thus radio astronomical survey work is easily understood as offering a unique contribution. Likewise, preceding and during the Second World War, the longitudinal and geomagnetic differences between Australia and other continents would prove especially interesting and even vital to study for ionospheric and radio propagation purposes.[13] Australia's varied weather conditions made atmospheric static ('strays') known as a problem from the earliest days of radio. Study of this would continue to be important through the Second World War.[14]

One of the first studies the RRB recommended in 1927 was to repeat 'Professor Appleton's Experiments'.[15] Also, the government needed to know what frequencies to use to communicate with the population living in the remote areas of Australia. In the spring of 1929, the RRB advertised for radio scientists. Madsen set up an English panel (Rutherford, Henry Tizard and Appleton) to vet British candidates. There was one Australian candidate, who withdrew, and eleven foreign candidates of whom the panel recommended six. The RRB chose four: Drs Leonard G.H. Huxley and David F. Martyn, and Messrs. A.L. Green and George H. Munro. There were considerable difficulties in getting the candidates to come and the RRB was assisted by the DSIR in England and by the General Electric Company in the USA with training and information.[16] Soon after they arrived, J.L. Pawsey did his M.Sc. work in Laby's laboratory at Melbourne, working with Munro and Huxley on atmospherics. W.G. Baker, O.O. Pulley and Huxley worked on radio field strength measurements out of the Russell Engineering School of the University of Sydney under Madsen, using equipment modelled after that designed by Harald T. Friis and E. Bruce at Bell Laboratories in the United States.[17] But the most exciting work done in the early ionosphere days in Australia was A.L. Green's polarization tests. Green had studied with Appleton at King's College London. Appleton and Ratcliffe (and Green!) had studied the polarization of downcoming waves from the ionosphere in England and found them to be elliptically polarized, usually in the left-handed sense.[18] Now, a test of the newly established magneto-ionic theory should show that if the 'electrical carriers in the ionized layer are of electronic mass, ... similar measurements made in the Southern Hemisphere' should show right-handed polarization. Appleton urged Green to take the RRB job and to do ionospheric layer measurements in Sydney.[19] Green arrived in Sydney in December 1929. After a great deal of difficulty getting an appropriate site and his equipment working, Green telegraphed Appleton in September 1931 that he had succeeded in showing that downcoming waves in the southern hemisphere were right-hand polarized. Appleton found this very exciting.[20] Green did, too, and submitted a paper to the Royal Society. Unfortunately, Green did not succeed in getting this paper published, but Appleton published an article concerning it in *Nature* in December 1931. Green finally succeeded in publishing his findings in Australia in 1932 and in the United States in 1934. Green pursued his ionospheric work with the RRB until joining AWA in 1935, where he continued radio and ionospheric work and eventually became editor of the *AWA Technical Review*.[21] D.F. Martyn showed *his* independence by sending off an ionospheric paper to *Nature* without getting permission from anyone.

The economic depression of the 1930s hurt the RRB. Though it was giving research grants to Melbourne and Sydney, its 'officers' working for the RRB (Green, Huxley, Martyn and Munro) were in a poor situation. Baker left to join AWA in July 1931 and Huxley left for England in September of the same year. Laby and Munro did not get along and so Munro moved to Sydney, followed by Martyn. R.O. Cherry resigned to go to England. This opened up positions for native Australians Geoffrey Builder and Hugh Webster to return from England and join the RRB.[22] Builder, and after him O.O. Pulley, did very creative instrumental work on the ionosphere. Builder was one of the reasons for Appleton's ionospheric success at Tromso, Norway, during the Second Polar Year and was the major contributor to helping Appleton change to the pulse method of ionospheric sounding. Builder left the RRB but continued designing radio equipment and publishing about this and worked on radar during the War, later taking up relativity as a research interest. It can be argued that Baker, Huxley, Cherry, Builder, Green, and Webster left the RRB at least partly for financial reasons during the Depression. A similar thing happened in other countries; for example, Lloyd V. Berkner was released from the US National Bureau of Standards and found a home at the Carnegie Institution's Department of Terrestrial Magnetism during the Depression when the Bureau was forced to reduce its staff by one quarter on very short notice.

The greatest contributor to Australian work in ionospheric and radio physics was D.F. Martyn.[23] He came with a PhD from London, worked with Laby's group in Melbourne for two years on signal fading and ionospheric layer heights, and then moved to Sydney. Martyn was among the most prolific of writers. In fact, during the period 1925–1960, Martyn produced at least 35 ionospheric papers, fourth highest among a sample of 1,676 authors publishing in ionospheric physics.[24] In a survey conducted in 1972 which received responses from more than 450 persons who had contributed to ionospheric physics, Martyn was mentioned fourth most frequently (after Appleton, Chapman and Ratcliffe) as an important founder of ionospheric physics, and he was mentioned fifth most frequently as someone influencing *current* (1972) research, two years after his death. Other Australian 'founders' mentioned in the survey were J.L. Pawsey, V.A. Bailey and G.H. Munro.[25]

The format of this chapter does not allow a detailed presentation of the research contributions of a number of Australian radio physicists. The following notes concerning D.F. Martyn would, however, seem to be appropriate since he is generally regarded as Australia's outstanding physicist from the period between the World Wars. In 1930 Martyn urged VHF radio experiments to reflect radio waves from the surface of the Moon. In 1933 he collaborated with V.A. Bailey in studying the so-called 'Luxemburg Effect', a non-linear modulation of one ionospheric wave by another. These 'heating' effects stimulated Bailey throughout a long career to study means of artificially modifying the upper atmosphere, something that has become an active field of research since the late 1950s. In 1935 Martyn designed a Doppler radar that Pulley built. Although this was not successful, it was a forerunner to a radar system widely used today. Martyn's radar device was designed to measure horizontal movements in the ionosphere. Such movements were in fact detected in those years by Munro, although he did not pursue the studies more directly until after the War. In 1935 Martyn used conformal representation to obtain dispersion and absorption curves for magneto-ionic propagation and worked with Bailey on this.

In the mid-1930s, a key problem was whether the free electrons in the ionosphere disappeared through recombination or through attachment. This was taken up in 1936 by Martyn and Pulley in a highly original paper in which various aspects of the ionosphere were discussed; it was, for example, deduced that the temperature in the F-region of the ionosphere is around 1,200°K.[26] In 1938 Martyn and Munro argued (correctly) against many others that the Lorentz polarization term in magneto-ionic theory should be equal to zero.[27] From 1936 or so, Martyn turned to the relations between the ionosphere, geomagnetism and solar physics. Much of his research was centred on the F-region. Conclusions that he reached were negative because of his failure to appreciate that the virtual heights measured by ionosonde recorders are not true heights. His recognition of this long-continued failure was in part responsible for his tragic suicide in 1970.[28] For a portion of his career Martyn was closely tied to the Mt Stromlo Observatory. He was clearly recognized as Australia's expert on the ionosphere. In 1936 he won a leave and visited the United States and Europe, representing Australian physics at several scientific meetings. Again, in 1939, just before the outbreak of war in Europe, Martyn visited England to learn about radar on behalf of Australia. Upon his return, he and others resigned from the RRB to join the secret work of the newly established Radiophysics Laboratory.[29]

Martyn had some personality difficulties with old colleagues as the years went by, but he increased his international stature as a radiophysicist and spokesman for Australian physics. He was elected FRS in 1950 and helped found the Australian Academy of Science in 1954.[30] He was named chairman in 1950 of the newly established Commission V on Radio Astronomy of the URSI. In 1954 he moved to the chairmanship of URSI Commission III on the Ionosphere, replacing Appleton who was by then a Nobel Laureate. It was Martyn who urged and arranged for URSI (The second oldest international scientific body) to hold its first general assembly in the southern hemisphere, in Sydney, 11–23 August 1952. The radio astronomy session had nineteen papers, with only two from the United States, the rest being from Australia, the Netherlands, England and France. Martyn was also named to the URSI Commission on the IGY.

Finally, a comment on the stature of Martyn as a person. Australians in 1950 were very bitter about Japanese actions towards Australian prisoners during the Second World War.[31] It was decided not to invite Japan or Germany to be part of the Xth General Assembly of URSI in Sydney. Martyn's knowledge of Japanese work on disturbance and motions in the ionosphere and his own work in these areas placed him in a position of high esteem among the Japanese. It was agreed that following the URSI meeting in Sydney, Martyn would visit Japan to transmit the news of URSI happenings.

A number of Australian radio and ionospheric scientists received graduate training or professional experience in England. Victor A. Bailey received the first PhD from Oxford University in 1924 and came out to Australia.[32] Laby had studied at Cambridge. Leonard G.H. Huxley studied at Oxford with J.S. Townsend. The New Zealander G.H. Munro worked at the Radio Research Station, Slough, before moving to Australia. The New Zealander Miles A.F. Barnett, the Australians Geoffrey Builder, O.O. Pulley, Frank Wood and J.H. Piddington; and others who moved to Australia such as A.L. Green, F.W.G. White and E.G. Bowen studied with Appleton, either at Cambridge or at King's College in London. The Tasmanian Hugh C. Webster studied at Cambridge with Rutherford. The tradition of radio study in England continued after the Second World War, although the first *two* generations of Australian radio physics teachers (Madsen, Laby; Webster, Huxley) were building up a solid core of Australian-trained men, at least through the MSc level. J.A. Ratcliffe at the Cavendish Laboratory trained more eminent ionospheric and radio workers than anyone in the world in the period 1925–1960.[33] Thus, Ronald N. Bracewell and Francis F. Gardner (native Australians), and Basil H. Briggs, studied with Ratcliffe at the Cavendish just after the war. Kevin C. Westfold studied with Sydney Chapman at Oxford, and John Davis with J.G. Davies at Manchester.

We find that the development of ionospheric physics, particularly at the Universities of Queensland (Webster) and Adelaide (Huxley), resulted in training a good number of radiophysicists in the two decades after the war. For example, of 27 persons sampled in 1972 who had done at least some ionospheric physics during their careers, had done most or all of their ionospheric research in Australia, and had done a research thesis, Webster had directed seven research theses, Huxley four, Ratcliffe (Cavendish) three, Laby two and eleven other persons one thesis each. Thus for the sampled group, whose research theses were completed over a period of almost four decades, fifteen persons had directed a total of 27 theses. Eleven of those fifteen persons had directed the theses while in Australia or New Zealand; a total of 21 of the 27 theses were directed in Australia (19) or New Zealand (2).[34] Ionospheric work during the Second World War and the establishment of working ionospheric stations during this period seem to have helped these Australian research groups develop and to have assisted them to continue turning out good research students. This is somewhat similar to the case in other countries, where Stanford (California), and Manchester (England) built up radio groups using ionospheric, radio and radar equipment with people trained during the war. Also, when Australian universities were enabled to grant the PhD degree several years after the war, they found it easier to keep research students.

Though seen as a national rather than a predominantly international one, the outstanding Australian journal for radio and ionospheric publications for the period from its founding in March 1935 until after the Second World War was the *AWA Technical Review*, a quarterly published by Amalgamated Wireless (Australasia) Ltd. Its editor from 1935 until 1941 was W.G. Baker, a key figure in Australian radio and ionospheric research who had earlier worked for a time at General Electric's radio laboratories in Schenectedy, New York. He was succeeded as editor by A.L. Green, mentioned above. Baker had strong ties to the American Institute of Radio Engineers, whose publication *Proceedings, Institute of Radio Engineers*, had been the premier radio engineering journal since the time of the First World War.[35] Other radio publications utilized by Australians were those of the Institution of Electrical Engineers in England and the Institution of Engineers, Australia. Papers of a more physical orientation were submitted to the *Proceedings of the Physical Society of London*, to the publications of the Royal Society of London, and to *Nature*. In the early years, papers were often published in Australia in the *CSIR Bulletin* (being issued separately as Radio Research Board *Reports*). The outstanding Australian physics publication in the years following the Second World War was the *Australian Journal of Physics*, which evolved from earlier publications and was founded in 1953.[36] This journal was the twelfth most cited journal in a study of ionospheric physics literature citations for 1970. It had an 'impact ratio (that is, the ratio of cited to citing references) of 1.8 for the 1,891 authors in the sample studied, one of the highest ratios of any journal used in ionospheric physics.[37] Still, Australian ionospheric physics authors follow the pattern today of placing their most important papers in the leading international specialty journals within a given discipline, rather than within a national society journal.

At the outbreak of war, Australia's economy was unevenly prepared. Professor Madsen had visited the United States National Bureau of Standards and England's National Physical Laboratory and had been trying to establish a similar organization in Australia for some years. Just as war began, the Australian National Standards Laboratory was founded upon the action of George Julius, chairman of CSIR and also chairman of a government-appointed committee charged with recommending changes in Australian industrial testing and research.[38] There were then three key Australian companies in radio assembly and manufacturing: AWA, which had manufactured electronic tubes since 1931, Phillips Electrical Industries of Australia Pty. Ltd. and Standard Telephones and Cables Pty.

Ltd.[39] With many Australian males in military service, women such as Florence Violet McKenzie took responsible roles in radio training, operating and construction.[40]

At the beginning of the war, most Australian radio and ionospheric researchers were at Sydney or Melbourne Universities, at the Radio Research Board, or in radio companies such as AWA. On 24 February 1939, Australian Prime Minister Lyons received a telegraphed request from London to send the 'best qualified physicist' to England. George Julius and Madsen recommended D.F. Martyn.[41] Martyn returned from England in August 1939 educated about the history and present state of radar work. Martyn had already gathered more or less what was being done in England from his visit there in 1936.[42] The government agreed to set up a classified radar research laboratory, the Radiophysics Laboratory, founded in Sydney with Martyn as its first Chief. Thus, Australian Radio and ionospheric research during the Second World War would be channelled into two organizations: the Radiophysics Laboratory and the Ionospheric Prediction Service of the radio Research Board.[43] The RRB lost a good part of its senior and best people to the Radiophysics Laboratory, which grew to a staff of over 200, including 65 scientists. For example, Munro, J.H. Piddington, F.J. Kerr and others soon joined Martyn at the Laboratory. Eventually, this led in 1941 to the closing of Laby's RRB laboratory at the University of Melbourne. There was a considerable amount of bad feeling about this at the time, but in truth the work at Sydney had outstripped that at Melbourne for some years.[44]

One of the early promises of ionospheric research was that it would be possible to choose the frequencies, times and directions of transmissions for optimum radio communication. Early 'soundings' of the ionosphere were taken manually, and only two to four times during a day or even less often. In the mid-1930s there were only three or four locations in the world making any sort of regular measuring runs, and even by 1939 there were only seven. The first automatic multi-frequency ionospheric sounder was built by T.R. Gilliland at the US National Bureau of Standards and went into operation on 20 April 1933. Robert Naismith at the NPL at Slough, England, was another pioneer, as was Lloyd V. Berkner at the Department of Terrestrial Magnetism (DTM) of the Carnegie Institution of Washington. The first ionospheric soundings in the southern hemisphere were performed by A.L. Green in 1931.

Berkner and the DTM played a big part in the history of geophysics and radio science in Australia. The DTM was founded in 1904 to carry out a world-wide magnetic survey, complementary to the Carnegie Institution's series of oceanographic world surveys. Douglas Mawson, Kerr Grant and others had long been interested in geophysical, magnetic and radio studies in Australia and in the Antarctic.[45] Thus the DTM was welcomed when it installed a geomagnetic observatory at Watheroo, Western Australia. This was practically the only physical research site in Western Australia at the time. By 1934, the DTM station at Watheroo was performing radio atmospherics, geomagnetic and solar measurements, and was installing a semi-automatic ionospheric sounder. Shortly thereafter, the DTM developed its own automatic multi-frequency ionospheric sounder and installed one at Washington, DC and one near the geomagnetic equator in Huancayo, Peru. The DTM hoped to place another in Alaska, and one at Watheroo. Lloyd Berkner had previously conducted radio propagation studies in New Zealand during 1929–1930 while a member of Admiral Byrd's First Antarctic Expedition.[46] Berkner visited Australia in 1938–1939 and installed the new automatic ionospheric sounder at Watheroo. This tradition of co-operation with DTM led to the Australian Ionospheric Prediction Service in the Second World War for, during his visit, Berkner and D.F. Martyn signed a report outlining such plans for co-operation between the DTM and the RRB. By 1942, in fact, all RRB ionospheric sounding was being done by ex-Watheroo men.[47]

During the mid-to-late 1930s, systems of propagation forecasting were being developed. Newbern Smith at the US Bureau of Standards, Thomas L. Eckersley at British Marconi, and E.V. Appleton and W.G. Beynon at Slough, each had distinctive methods. The main ones were Smith's 'transmission curve' approach and Appleton's 'parabolic layer' model.[48] The US NBS had begun sending radio propagation forecasts world-wide in 1935 over frequency-standard radio station WWV. F.W. Wood worked with the vastly undermanned RRB to develop ionospheric predictions for the Australian services, particularly the military and the government's civilian short-wave service. By 1940–41, Wood was receiving ionospheric data from Madras, India; New Zealand; Washington, DC; Huancayo, Peru; and England. He would later receive data from US stations at Cambridge, Massachusetts; Stanford, California; Baton Rouge, Louisiana; and Puerto Rico. His network in Australia consisted of sounders at Watheroo, the Mt Stromlo Observatory and Sydney, and he had several years of Australian data. New Zealand radio and ionospheric workers about 1939 included P.W. Burbidge, D.C.H. Florance, R. Jack, M.A.F. Barnett, G. Searle, T.W. Straker, K. Kreielsheimer and F.W.G.

White.[49] White came to Sydney in 1941 to take up a senior position at CSIR and the Radiophysics Laboratory. White and Wood had both worked with Appleton in England and White was of immense help in pushing through support for Wood to build up the Ionospheric Prediction Service for Australia.[50]

The radio prediction system, to be effective, necessitated a complicated international operation that called for more and more co-operation as the war progressed. Australia would find itself playing a central part in this system, given the importance of military campaigns in the Asian, Indian and Pacific areas. In 1941, F.W.G. White, E.V. Appleton and J.H. Dellinger (Director of the US NBS Radio Laboratories), simultaneously suggested the need for better Allied co-operation in radio forecasting. More and more ionospheric stations were added during the war (eventually including ones at Brisbane, Cape York, Port Moresby, and various Pacific islands). Still, the Allies encountered problems in forecasting – geophysical and human problems – and these led to a major radio propagation conference of the Allies held in Washington in the Spring of 1944. Operating in the jungle took a heavy toll on radio equipment. Australian experts became world authorities on 'tropic proofing' equipment. But heavy jungle also made radio operations nearly impossible – signals didn't get through. Different services within a given country felt they had different needs and priorities. Services of different countries had the same problems. Some (for example the armies) needed short-range forecasting accuracy. Others (the navies) needed forecasting in areas, such as vast ocean surfaces, for which there was little applicable data. The beam broadcasting services between North America and the British Empire needed very long-distance forecasts.

Four networks were used by the Allies. The Americans established a combined military-civilian propagation laboratory at the Bureau of Standards in Washington, the Interservice Radio Propagation Laboratory (IRPL). This group used Newbern Smith's transmission curve method of forecasting. The British had two somewhat antagonistic groups. T.L. Eckersley's group, the Inter-Service Ionospheric Bureau (ISIB), operated for the British military and co-operated closely with the Americans. Eckersley's forecasting model approximated that of Newbern Smith. Appleton and Eckersley had not personally got along since about 1930. Appleton's group at the Radio Research Station, Slough, worked for British civil radio. They used the Appleton-Beynon parabolic layer model. The Australian service originally used Smith's curves, then changed to a version of their own, using Appleton's model. It was claimed that there were differences of 20% to 50% in the values of the predicted optimum working frequencies, depending upon the forecasting model chosen. The military services argued that simple tables should be distributed to operators. Handbooks had been quickly written in 1942 and brief ionospheric mini-courses were inserted into the military radio training courses.

A good part of the problem was due to the asymmetry and displacement of the Earth's geographic coordinates and the geomagnetic field.[51] Basically, the Earth's ionized layers are partly controlled by solar radiation and its interaction with the gases in the upper atmosphere. Thus, geographic latitude and longitude play a role. But the geomagnetic field also plays a part in ionospheric currents and density distribution. The geomagnetic field is skewed so that some equatorial latitudes are more like some mid-latitudes. In other words, there is a 'longitude effect' to the ionization of the upper F_2 region of the ionosphere. Failure to recognize and allow for this leads to significant errors (30% to 100%) in propagation forecasting. At the International Radio Propagation Conference in Washington in April 1944, the longitude effect was discussed by all.[52] Later, Appleton (who as head of the British DSIR could declassify materials for publication) first published in the open literature a note in *Nature* concerning the 'longitude effect' and claimed priority for a major discovery.[53] A good deal of unhappiness was generated by this action, for Appleton was not the first, but the third or fourth, to discover and write about this effect! The discovery was made by K. Maeda, H. Uyeda and H. Shinkawa and published in Japanese in April 1942.[54] It was next discovered in the early winter of 1943 by M.L. Phillips, N. Smith and D.K. Bailey and included in the IRPL forecasts issued in summer 1943. T.L. Eckersley's English ISIB group knew of it also in 1943 and discussed it in correspondence. Finally, in December 1941 F.W.G. White and F.W. Wood had discussed with the Australian military the fact that data from Huancayo, Peru, was not good for Port Moresby forecasts due to the big longitude difference, although they may not at that time have understood the reason why.

In any event, the Washington conference solved a lot of problems. The solution to successful jungle radio transmissions was to get the antennas above the jungle cover and utilize sky-wave rather than ground-wave. The longitude effect would be studied by utilizing new stations being put into operation in 1944 and by utilizing the corrected tables issued by groups such as IRPL. Atmospheric noise was a great problem in the Australian area of operations in the south-west Pacific and would be studied further. Supposedly slight changes made to the application of the various models reduced the difference in forecast results to a mere 2 to 3%. New handbooks and improved training methods were planned. Moreover, the British would co-ordinate their two competing forecasting services (Appleton's and the ISIB) as the ISIB, US-IRPL, and Australian groups had been doing since 1942. The rest of the British Empire also was included in this overall plan, as were Russia and China.

Australia was represented at these meetings by George Munro, the Australian scientific representative in Washington, by F.W. Wood, representing the RRB, by S/L A.L. Hall of the RAAF, and by Lt. A.W. Moriarty of the RAN. The New Zealand government was represented by C. Ellyett and G. Searle, by their representative in Washington, K.D. George, by S/L A.W. Stockwell of the RNZAF, and by Commander B.J. Giles of the RNZN. This conference was more than just a two-week get-together during the latter part of the Second World War. The 1944 Washington radio conference had been in preparation for months and was the first major international meeting to co-ordinate radio forecasting. From Australia's standpoint, it acknowledged Australia's geophysically and geographically important position in world-wide radio communications, and gave American-Australian relations, which had grown close during the joint military operations in the Pacific, a further base upon which to build in the post-war world.

Ionospheric radio development continued following the war. The Ionospheric Prediction Service had become a generally useful synoptic service, and it was decided to make it a separate unit under the Commonwealth Observatory with A.L. Green as Officer-in-charge. D.F. Martyn and G.H. Munro returned to the Radio Research Board, with Martyn as Chief Scientific Officer remaining at Mt Stromlo and Munro as Officer-in-charge of the Sydney Section, located at the Electrical Engineering Department of the University of Sydney. The RRB assisted radio groups to begin or to develop further at the Universities of Queensland, Western Australia, Adelaide and Tasmania.[55] These groups, particularly those of Hugh C. Webster in Brisbane and Leonard G.H. Huxley at Adelaide produced numerous good scientists. Adelaide and Queensland continued strong in the 1960s and 1970s with leaders such as B.H. Briggs and J.D. Whitehead. K.D. Cole, who trained at Queensland, became a major figure in the 1960s at La Trobe University. Others, such as the Queensland product J.A. Thomas, taught and did research at the RAAF Academy. Still others began teaching ionospheric physics and related studies at the Australian National University, the University of Newcastle, or the University of New England, or they took research positions at CSIRO (formerly called CSIR) or at the Weapons Research Establishment.

Significant research projects were undertaken with grants and contracts from the United States beginning in the early 1960s. Research topics reflected international trends toward magnetospheric physics, space physics, remote sensing, and over-the-horizon ionospheric radar.[56] The Australian and New Zealand interest in polar upper atmospheric physics and Antarctica increased in the 1950s and continues to the present time. There has been particularly warm co-operation between New Zealand, Australia and the United States during the IGY and in post-IGY research in Antarctica.[57]

Radio astronomy, particularly in the quarter century following the Second World War, has been one of the most outstanding areas of Australian achievement and its history has been extensively investigated.[58] Native and transplanted Australians became renowned for studying at radio wavelengths the Sun, planets, galaxies and beyond. It was *radio* science, not classical astronomy, that welcomed the radio astronomers, and so it was especially significant that D.F. Martyn persuaded URSI to meet in Sydney in 1952 for its Xth General Assembly. Following the First World War, the United States built and maintained a world position in ionospheric physics and radio engineering but, following the Second World War, military opportunities for ionospheric research and opportunities in nuclear and high-energy physics perhaps kept most American physicists and radio scientists away from radio astronomy. So when the United States entered the field a decade late, Australia had outstanding radio physicists to send to direct its new radio astronomy observatories and to lead American research: J.G. Bolton and G.J. Stanley at Cal Tech, Ron Bracewell at Stanford, Frank J. Kerr at Maryland. J.L. Pawsey's untimely death in 1961 prevented him from taking up the position of head of the new US National Radio Astronomy Observatory (NRAO). Here, there is also a tie back to Pawsey's friend Lloyd Berkner and earlier ionospheric days in Australia. In the late 1950s and the 1960s, Berkner was president of Associated Universities Incorporated, the organization designing the US NRAO at Green Bank, West Virginia.

The growth of ionosphere and radio studies in Australia can be discussed also in a scientometric sense by examining statistics and demographics. The late Derek J. de S. Price estimated that science as a whole in this century has grown approximately at an exponential rate of about 5% per year, or that science doubles in size about every fifteen years.[59] From L.A. Manning's carefully prepared bibliography and other data, it has been determined that the literature of ionospheric physics as a whole grew at the very high rate of 21% per year (doubling in size every 3.2 years) for the inter-war period 1926–1938, and grew at a slightly lower but still very high rate of 13% per year for the period 1947–1969 (doubling in size every 5.4 years).[60] Thus ionospheric physics was a vigorously growing transitional discipline, growing much faster than science at large. (However, Ellyett has noted a decline in growth since the 1960s.)[61] It is difficult to measure the growth rate of the literature in individual countries before the time of the Second World War, since the habit of giving authors' addresses on papers became the norm only after 1945.[62] But for the period approximately 1947–1960, Australian, Indian, Japanese and Soviet ionospheric physics literature grew at a doubling rate of 3 to 4 years, faster than the general ionospheric literature. For the period 1925–1960, Australia was the fourth-ranking country (after the United States, England and Japan), while New Zealand was the tenth-ranking in the total number of ionospheric physics papers produced, among some 45 countries producing at least one article during that period. Of the top one hundred most productive authors of ionospheric papers (out of 1,676), fifteen had done most or all of their ionospheric physics research in Australia.

In a 1972 mail survey of persons who had published at least one paper of interest to ionospheric physics, more than 450 responses were received from 31 countries.[63] For analytical purposes these were grouped into three world regions: 'USA', 'Asian-Pacific' (Australia, Ceylon, Hong Kong, India, Japan, New Zealand and Taiwan), and 'Europe and Other' (23 countries in Europe, Africa, South America and Canada, but with most responses from Europe and Canada).[64] The study found some correlation between research training and world-region activity in ionospheric physics. The United States has a smaller proportion of physicists who have done ionospheric physics than the other two regions, but a higher proportion of astronomers who have done some ionospheric physics. 'Asia-Pacific' has the lowest proportion of engineers, and 'Europe and Other' the lowest proportion of geophysicists. For the nine largest degree-granting countries in the study, only 8% of the sample were doing engineering in 1972, though 20% of all degrees earned were engineering degrees. And conversely, though 20% were doing astronomy in 1972, only 6% of the earned degrees were in astronomy. This indicates that a number of persons trained in engineering or in geophysics shifted during their careers to (radio) astronomy and ionospheric physics. Results also reflect the disciplinary traditions of different countries: Until the Second World War, Japan operated under the German model where much early ionospheric work was done by people trained in geophysics and geomagnetism. After the war, Japan adopted the American model of doing ionospheric physics under the umbrella of Electrical Engineering. Similarly, Australia, Canada and New Zealand somewhat followed the model of England, where ionospheric and other physics was usually done in physics and mathematics departments.

The study also investigated geographical mobility among those who have done ionospheric physics. Comparing the country of birth of each individual to the country of employment in 1972 for nine countries, Canada and the United States had the highest proportion of immigrants, while West Germany, India and England had had the most emigration. Australia and New Zealand had almost zero net emigration. The median age of the total responding sample was 48 years (birth year 1924), with a very large variance. In a related study, six countries in the survey sample (USA, Japan, New Zealand, England, India and Australia) were examined for evidence of travel abroad for purposes of study or research in ionospheric physics. Among those who had never had significant (defined as at least six months) research or study periods in another country, the United States was the highest 'stay-at-home' with 71%, Australia the least with 39%. Thirty-six percent of the American sample were immigrants, most having already received their ionospheric training abroad. England had the greatest 'outflow' or emigration of ionospherically trained workers. This does not mean that England had the highest percentage of native-born workers emigrate, but measures in part the fact that for decades, England was an educator of students from the Empire. Of those in the survey who did their first ionospheric research in England, 24% subsequently emigrated. The USA and Canada were the large importers of trained ionospheric workers, while England and Germany, proportionally, were exporters. Australia stands out as a country with a considerable amount of geographical mobility, with some immigrants balanced by a good number of emigrants, and lots of Australians who did research and training for a time abroad. In the sample, the countries of largest flow to and from Australia were the United States and England.

Another measurement in the 1972 survey concerned the social mobility of those in the sample. A question asked, 'What kind of work did your father do when you were about sixteen years old?' This is a standard type of question in social surveys and is used to measure the family situation at the time when a youth is about to enter the job market or higher education. Respondents in the table were placed in one of the three world-regions by their country of birth. The total Asian-Pacific sub-sample exhibited less upward social mobility. That is, a larger proportion of ionospheric workers born in the United States were sons of skilled, semi-skilled and unskilled workers than was the case with the other groups. (Only 2% of the 1972 survey respondents were female.) However, the combined Australian and New Zealand sub-samples (n=40) segregated from the Asian-Pacific group (n=94) seem to compare more closely to the USA group than to the Asian-Pacific one. Thus, given the limitations of generalizing from this rather small sample of about 400 respondents and the even smaller sub-samples within it, the United States and Australia seem to have been countries of more than average professional social opportunity for those who would make their careers in ionospheric and related areas of physics.

This paper has told a story of Australia's ties to England, its increasing ties to the United States, and its own creation and innovation in ionospheric and radio studies from the period before the First World War until the age of satellite communications in the 1960s and 1970s, and has placed these developments upon a more general background of ionospheric and radio research in other countries. The creation and evolution of the Australian Radio Research Board and its relations with university and industrial research formed the most central force in Australian radio science until the Second World War. A general sketch of the backgrounds and education of Australian researchers has been given with the careers of several persons discussed in some detail, and with some prosopographical data. The major university tradition initially was to gain advanced research knowledge in England. This later changed somewhat as Australia's universities – Melbourne and Sydney, and later Brisbane, Adelaide and others – built and enlarged their own traditions. Some of the tools of research have been examined – the experimental equipment and methods, the theoretical ideas, and the modes of communication. Australians made world-recognized contributions to ionospheric physics as early as a decade before the Second World War, but some of their studies (examples given here mentioned Martyn and Munro) suffered from delay and isolation due to Australia's location. But location on the 'underside' of the globe would prove especially interesting in, for example, the ionospheric geomagnetic longitude anomaly, and in numerous areas of radio astronomy. The increasing relations with the United States during the war in the fields of radio and geophysics had been presaged by the pre-war co-operation with the Department of Terrestrial Magnetism and the National Bureau of Standards. The development of Australian, and Allied, ionospheric prediction techniques represented a merger of world-wide upper atmospheric geophysics and solar physics research into war-related work. The 1944 Washington Allied Radio Conference represented an important occasion not only for global communications policy and methods but also for the recognition of Australia's important geographical and geophysical position in the world. The transformation of Australian radio and ionospheric research from wartime to peacetime activities and the new movements in post-war research truly took Australia from isolation into full global membership among the leaders in radio science and radio astronomy and into the arena of 'Big Science'.

Department of History
Wesleyan University
USA

NOTES

1 C.S. Gillmor, 'Wilhelm Altar, Edward Appleton, and the Magneto-Ionic Theory', *Proceedings of the American Philosophical Society*, 126 (1982), 395–440. C.S. Gillmor, 'The History of the Term "ionosphere"', *Nature, 262* (1976), 347–348. C.S. Gillmor, 'Threshold to Space: Early Studies of the Ionosphere', in *Space Science Comes of Age*, ed. Paul A. Hanle and Von Del Chamberlain (Washington, DC 1981), 101–114. C.S. Gillmor, 'S.K. Mitra's "The Upper Atmosphere": The Role of Monograph and Text Literature in the Evolution of Ionospheric Physics', *Indian Journal of Radio and Space Physics*, 15 (1986), 171–181. Also: A.L. Green, 'Early History of the Ionosphere', *AWA Technical Review*, 7 (1946), 177–228. H.R. Mimno, 'The Physics of the Ionosphere', *Reviews of Modern Physics*, 9 (1938), 1–43. J.A. Ratcliffe, ed., 'Special Issue: Fifty Years of the Ionosphere', *Journal of Atmospheric and Terrestrial Physics*, 36 (1974), 2069–2319.

2 C.S. Gillmor, 'Federal Funding and Knowledge Growth in Ionospheric Physics, 1945–81', *Social Studies of Science*, 16 (1986), 105–133.

3 R.W. Home, 'The Beginnings of an Australian Physics Community', in Nathan

Reingold and Marc Rothenberg, eds., *Scientific Colonialism* (Washington, DC, 1987), 3–34. R.W. Home, 'Origins of the Australian Physics Community', *Historical Studies,* 20 (1983), 383–400. R.W. Home, 'W.H. Bragg and J.P.V. Madsen: Collaboration and Correspondence, 1905–1911', *Historical Records of Australian Science,* 5 (1981), 1–11. Also: Ann Moyal, 'The History of Telecommunication in Australia: Aspects of the Technological Experience, 1854–1930', in Reingold and Rothenberg, op. cit., 35–53. Ann (Mozley) Moyal, *Clear Across Australia: A History of Telecommunications* (Melbourne, 1984). L.A. Hooke, 'Australian Radio Communication Services', *AWA Technical Review,* 3 (1938), 229–251.

[4] George H. Munro, interview with C.S. Gillmor, Sydney, 27 September 1973.

[5] F.W.G. White, 'Early Work in Australia, New Zealand and at the Halley Stewart Laboratory, London', *Phil. Trans. Roy. Soc. Lond., A.,* 280 (1975), 35–46. F.W.G. White; interview with C.S. Gillmor, London, 5 December 1974.

[6] George Currie and John Graham, *The Origins of CSIRO: Science and the Commonwealth Government, 1901–1926* (Melbourne, 1966). F.W.G. White, 'A Personal Account of the Historical Development of CSIRO', *Nature,* 261 (1976), 633–636.

[7] F.W.G. White, op. cit. (n. 5; 1975) p. 37.

[8] Much of the information that follows concerning Australian radio and ionosphere work draws upon W.F. Evans, *History of the Radio Research Board, 1926–1945* (Melbourne, 1973.

[9] C.S. Gillmor, op. cit. (n. 1; 1982), p. 397.

[10] See, for example, A.L. Green, op. cit. (n. 1).

[11] Notes concerning Geoffrey Builder are in the papers of E.V. Appleton, University of Edinburgh Library; at the Radio Research Station, Slough, England; and in materials provided me by R.W. Home. For a listing of Appleton's papers, see 'Catalogue of the papers and correspondence of Sir Edward Victor Appleton, G.B.E., K.C.B., F.R.S. (1892–1965), deposited in the Library of the University of Edinburgh' (Contemporary Scientific Archives Centre, 1981). For Slough, see the folder: D.S.I.R., Radio Research Station Library, 'International Polar Year. 1932. Related Papers. Packet I'.

[12] W.N. Christiansen, 'Australian Solar Radio Astronomy', and R.N. Bracewell, 'Imaging Theory Starting in Australia in the Fifties', both in W.T. Sullivan, III, ed., *The Early Years of Radio Astronomy* (Cambridge, 1984).

[13] Interservice Radio Propagation Laboratory (1944), 'International Radio Propagation Conference, held under auspices of Wave Propagation Committee, Combined Communications Board', at IRPL, National Bureau of Standards, Washington, DC, April 1944. Unpublished papers and minutes.

[14] Ibid.; also A.K. Cabrera, 'Atmospheric Noise, A Review of the Literature', *AWA Technical Review,* 12 (1962), 63–85.

[15] Evans, op. cit (n. 8), p. 20.

[16] Ibid., pp. 46–50.

[17] *CSIR Bulletin No. 47.* This is *Radio Research Board: Report No. 1* (Melbourne, 1931).

[18] E.V. Appleton and J.A. Ratcliffe, 'On a Method of Determining the State of Polarization of Downcoming Wireless Waves', *Proc. Roy. Soc. Lond., A,* 117 (1928), 576–588.

[19] C.S. Gillmor, op. cit. (n. 1; 1982).
[20] E.V. Appleton; Appleton letters to J.A. Ratcliffe, Edinburgh University Library, MS Gen. 1985. Letter of 30 September 1931.
[21] F.W.G. White, op. cit. (n. 5; 1975).
[22] W.F. Evans, op. cit. (n. 8), pp. 59–61, 83–84.
[23] H. Massey, 'David Forbes Martyn, 1906–1970', *Biog. Mem. of Fellows of the Roy. Soc.*, 17 (1971), 497–510. J.H. Piddington and M.L. Oliphant, 'David Forbes Martyn', *Records of the Australian Academy of Science, 2* (1972), 47–60. R.W. Home, 'David Forbes Martyn', *Dictionary of Scientific Biography, Supplement II* (forthcoming).
[24] L.A. Manning, *Bibliography of the Ionosphere* (Stanford, California, 1962).
[25] C.S. Gillmor, unpublished data.
[26] D.F. Martyn and O.O. Pulley, 'The Temperature and Constituents of the Upper Atmosphere', *Proc. Roy. Soc. Lond., A*, 154 (1936), 455–486.
[27] C.S. Gillmor, op. cit. (n. 1; 1982), pp. 414, 418.
[28] D.R. Bates, 'Recombination in the Normal E and F Layers of the Ionosphere', *Planet. Space Sci.*, 36 (1988), 55–63, and letter from D.R. Bates to C.S. Gillmor, 22 August 1988.
[29] W.F. Evans, op. cit. (n. 8), pp. 118–135, 236–7, 280, 290–291.
[30] D.F. Martyn, 'Personal Notes on the Early Days of Our Academy', *Records of the Australian Academy of Science,* 1 (1967), 52–72.
[31] John Robertson, *Australia at War, 1939–1945* (Melbourne, 1981) pp. 2, 86, 89, 206–207.
[32] Biographical and professional materials concerning V.A. Bailey provided through the courtesy of his daughter, Helen Bailey Bayly.
[33] C.S. Gillmor, unpublished data.
[34] Ibid.
[35] C.S. Gillmor and C.J. Terman, 'Communication Modes of Geophysics: The Case of Ionospheric Physics', *Eos,* 54 (1973), 900–908.
[36] B.J. Walby 'Australian Journals of Scientific Research', *Nature,* 261 (1976), 661–664.
[37] C.S. Gillmor and C.J. Terman, op. cit. (n. 35), p. 902.
[38] F.W.G. White, op. cit. (n. 6).
[39] S.J. Butlin and C.B. Schedvin, *War Economy, 1942–1945*: Volume IV of Series Four, 'Civil', in *Australia in the War of 1939–1945* (Canberra, 1977), pp. 403–406, 744, 767.
[40] Patsy Adam-Smith, *Australian Women at War* (Melbourne, 1984), pp. 10–11, 212. W.F. Evans, op. cit. (n. 8), pp. 331, 343–345. Mollie Bayne, ed., *Australian Women at War* (Melbourne, 1943).
[41] D.P. Mellor, *The Role of Science and Industry*: Volume V in Series Four, 'Civil', *Australia in the War of 1939–1945* (Canberra, 1958), p. 423.
[42] W.F. Evans, op. cit. (n. 8), p. 283.
[43] W.F. Evans' compilation and research on the history of Australian radio is contained not only in his 1973 volume, already cited, but also in a companion volume entitled *History of the Radiophysics Advisory Board, 1939–1945* (Melbourne, 1971).
[44] W.F. Evans, op. cit. (n. 8), pp. 291–293.
[45] C.S. Gillmor, 'Early History of Upper Atmospheric Physics Research in Ant-

arctica', pp. 236–262 in L.J. Lanzerotti and C.G. Park, eds., *Upper Atmosphere Research in Antarctica* (Antarctic Research Series Volume 29, American Geophysical Union) (Washington, DC, 1978).
46 Ibid., p. 241.
47 W.F. Evans, op. cit. (n. 8), p. 297.
48 C.S. Gillmor, op. cit. (n. 1; 1981).
49 W.F. Evans, op. cit. (n. 8), p. 151.
50 Ibid., pp. 321–323.
51 A classic description of this is given in S.K. Mitra, *The Upper Atmosphere*, 2nd ed. (Calcutta, 1952), Chapter VI, Section 11.
52 Unreferenced mention here and below of the International Radio Propagation Conference derives from unpublished records. See: Interservice Radio Propagation Laboratory (1944), op. cit. (n. 13). My copies of these were a gift from the conference recording secretary, the late physicist Marcella Lindemann Phillips.
53 E.V. Appleton, 'Two Anomalies in the Ionosphere', *Nature,* 157 (1946), 691.
54 K. Maeda, H. Uyeda and H. Shinkawa, 'Longitudinal Variation of the F_2-Region Electron Density (Differences of Ionospheric Conditions in the F_2-Region Between Two Points Lying on the Same Latitude But Having Different Longitudes)', *Rep. Phys. Inst. Radio Waves*, No. 2, pp. 1–13, April 1942 (in Japanese).
55 G.H. Munro, 'The Work of the Radio Research Board', *Proc. Inst. Radio Engineers, Australia,* 12 (1951), 41–43.
56 C.S. Gillmor, unpublished report, circulated 11 January 1974.
57 International Council of Scientific Unions, Scientific Committee on Antarctic Research (1966), *SCAR MANUAL* (Scott Polar Research Institute, Cambridge, England).
58 J.S. Hey, *The Evolution of Radio Astronomy* (New York, 1973). J.P. Wild, 'The Beginnings of Radio Astronomy in Australia', *Records of the Australian Academy of Science,* 2 (3) (1972), 52–61. W.T. Sullivan, III, ed., *The Early Years of Radio Astronomy* (Cambridge, 1984). E.G. Bowen, 'From Wartime Radar to Postwar Radio Astronomy in Australia', *Electrical and Electronics Engineering, Aust.,* 8 (1) (1988), 1–11. B.Y. Mills, 'Australian Contribution to the Science of Radioastronomy', *ibid.,* 12–23. W.T. Sullivan, III, 'Early Years of Australian Radio Astronomy', in R.W. Home, ed., *Australian Science in the Making* (Sydney, 1988), 308–344. See also R.N. Bracewell, op. cit. (n. 12) and W.N. Christiansen, op. cit. (n. 12).
59 D.J. de S. Price, *Little Science, Big Science* (New York, 1963).
60 L.A. Manning, op. cit. (n. 24). C.S. Gillmor and D. Gran, 'Research in Ionospheric Physics', *Actes* of the XIIIth International Congress of the History of Science, Moscow, 6 (1964), 160–164.
61 C.D. Ellyett, 'Worldwide Growth of Geophysics and Astrophysics Research Since 1967', *Eos,* 58 (1977), 124–126.
62 C.S. Gillmor, op. cit. (n. 2), p. 116.
63 C.S. Gillmor, unpublished report, circulated 11 January 1974.
64 C.S. Gillmor, 'Aspects of the History of Ionospheric Physics in the Asian-Pacific Area' (Abstract), *Proceedings, Second General Assembly, International Association of Geomagnetism and Aeronomy* (Kyoto, 1973).

H.E. LE GRAND

THEORIES OF THE EARTH AS SEEN FROM BELOW*

It is a commonplace in geologists' historical accounts that scientists in the southern hemisphere were much earlier in their acceptance of the idea of continental drift than their northern hemisphere counterparts. The explanation offered is captured in a pencilled scrawl by an anonymous wit on the flyleaf of a 1972 collection of papers by Australian earth scientists found in the library of the United States Geological Survey: 'Most of Earth's bigger scars, sutures and dimples etc. were more visible on her bottom, and the people down under were quicker to see them'.

If this common wisdom were well-founded, it could be accommodated within a familiar centre-periphery model of the international scientific community. One might claim that 'fringe' scientific views – as Drift was thought to be prior to the late 1960s – were more likely to be pursued on the geographical and intellectual 'fringe' (the periphery) of the scientific community where such views might escape the more rigorous scrutiny to which they would be subjected in Britain or North America (the centre). Alternatively, to pursue a theme of nationalism and internationalism in science, one might emphasize the socially contingent character of theory acceptance. One could construct an account of how there developed in Australia – through intellectual isolation or a concentration on 'local' scientific problems, or for whatever other plausible reasons might come to mind – a distinctive 'national' science favourable to Drift at a time when opinion in Britain and North America was decidedly unfavourable. This might even be the occasion for nationalistic breast-beating: the championing of Drift by southern hemisphere earth scientists, including Australians, signalled the end of the era in which Australians had quarried data that served as building blocks for American and European theorists; Australia now exported not only the raw materials (empirical data) but also manufactures (theories). Neither of these stories is the one I shall tell because, whatever the merits of these types of analysis, the common wisdom about the reception of Drift is not well-founded.

I shall indicate briefly in what ways the common wisdom is ill-

R.W. Home and S.G. Kohlstedt (eds.), International Science and National Scientific Identity, 205–232.

founded. I shall then sketch the theoretical and experimental work and institutional setting of a few earth scientists in Australia. My focus will be a dual one: Professor S. Warren Carey at the University of Tasmania, a prominent exponent first of Drift and then of an expanding earth; and the Department of Geophysics in the Research School of Physical Sciences at the Australian National University (ANU) which under the leadership of Professor John Conrad Jaeger in the 1950s and early 1960s rapidly moved to the forefront of selected areas of research, some of which, such as palaeomagnetics and rock dating, bore on Drift. I shall argue that within Australia, both of these 'institutions' were relegated to the periphery. Internationally, however, both received substantial recognition in the 1950s. I conclude by suggesting that in analysing the activities, interactions and views of the Earth of the ANU group and of Carey it is helpful to identify the scientific community in which they participated and from which they sought recognition and, within that community, modalities of interdependence. Finally, I return to the 'common wisdom' to consider what purpose it might serve for those who accept it.

I turn first to the assumptions embodied in what I have styled the common wisdom about attitudes toward Drift in the northern and southern hemispheres. One assumption is that there was a clear-cut difference in attitudes: north of the equator, a general and persistent hostility to Drift that did not soften until the late 1960s and early 1970s; south of the equator, an acceptance of Drift as the preferred global theory from the 1950s if not earlier. A second assumption, termed geological 'localism',[1] is that this difference is attributable to a relative abundance in the southern hemisphere of regional data and problems readily explicable in terms of Drift. Both of these assumptions are questionable.

First, it was not the case that northern hemisphere earth scientists unanimously opposed Drift from the 1930s into the mid-1960s. The response was more negative in North America than in Britain and in Britain than in some other European countries, but even in North America in the 1940s and 1950s – the nadir of Drift's fortunes – Drift as a global theory had not fallen into desuetude and reference was made to it as a legitimate if problematic and unpopular theory in many introductory geology texts.[2] It would be more accurate to say that at that time, in North America and perhaps in Britain, there was a lack of interest in Drift and indeed a lack of interest in most global theories. This was equally true in Australia. In Britain, and to a lesser extent in North America, there was by about 1960 an increased interest in and sympathy toward Drift. This was particularly associated with palaeomagnetic studies by scientists in Britain and elsewhere that pointed towards large-scale crustal mobility.[3]

Secondly, it is not obvious that the evidence of Drift was clearer in the south than in the north. The South African geologists Alexander du Toit in the 1920s and 1930s and Lester King in the 1950s amassed data in support of Wegener's claim that there was not only a physical fit but also numerous correspondences of geological structures between the opposing South Atlantic coasts. For Africa and South America, the 'jigsaw fit' was disputed by, among others, Harold Jeffreys, then doyen of British geophysicists and Plumian Professor of Astronomy at Cambridge. The arguments from jigsaw fit and geological pattern-matching were not, however, confined to the South Atlantic: Wegener and his supporters had made similar cases for the North Atlantic. For Australia, however, jigsaw fit and pattern-matching arguments could not yet be sustained. The major arguments for the inclusion of Australia in Wegener's giant continent Pangaea and for the subsequent separation and drift of Australia were usually taken not from geology *sensu strictu* but from species distribution and palaeoclimatology. Neither of these specialties was then in the mainstream of geology. Only those scientists in Australia who put primacy on solving Australian problems in these two fields were likely to be moved by Wegener's proffered solutions.[4] In any case, we should not overlook how evidence against Drift in the north was quietly transformed into evidence for Drift by the close of the revolution, as though it had always been there to see if only one had known where and how to look. It is wonderful how facts become so obvious once there is agreement on a preferred theory.

This leads us to a third objection to the common wisdom. It was not the case that southern hemisphere earth scientists overwhelmingly favoured Drift from the 1930s into the mid-1960s. In South Africa there were several partisans who gained national and international reputations for their advocacy of Drift. Recent studies[5] indicate that in Australia, as elsewhere, geologists were not overmuch concerned with theory, but of those who did express opinions of the rival global theories, only a very small minority preferred Drift. There had been a brief blooming of interest in Drift in the 1920s and 1930s among a few scientists in Australia interested in specific problems to which Drift promised a key. At the University of Sydney, for example, Launcelot Harrison, a zoologist, T.W.E. David, a geologist, and L.A. Cotton, a geophysicist, all evinced varying degrees of sympathy.[6] However, even in those years the general attitude was indifference mixed with scepticism and sometimes sarcasm. Drift, and indeed other global geological theories, seemed to offer little assistance in the work undertaken by most Australian geologists, many of whom espoused an atheoretical or even antitheoretical attitude towards grand global schemes.[7] By the early 1940s the Australian verdict on Drift was no less negative than that reached in North America; even those pursuing problems of Australian species distribution usually did not invoke Drift-type solutions.[8]

In the early 1950s, most members of the Australian geological professoriate were indifferent to Drift. Those who took notice of it were mostly antipathetic. C.E. Marshall at the University of Sydney, A.R. Alderman at the University of Adelaide and R.T. Prider at the University of Western Australia were petrologists; W.H. Bryan at the University of Queensland and A.H. Voisey at the University of New England focussed on regional problems to which Drift and other global theories seemed of little relevance. E.S. Hills strongly opposed Drift and recalled that during his long tenure at the University of Melbourne, Drift might have been mentioned in the staff tea room but that no one took it seriously.[9] K.E. Bullen, a disciple of Jeffreys, an internationally recognized seismologist and Professor of Applied Mathematics at the University of Sydney, was contemptuous of Drift. As late as 1975, in his last book, he gave only one passing reference to the revolution in earth sciences in which some of his fellow seismologists in North America had been active participants.[10] W.R. Browne, though not a member of the professoriate (he had retired in 1949 from his readership at Sydney), was a senior and very influential member of the Australian community. In his monumental edition of David's *Geology of Australia* he mentioned Drift only briefly and that was to conclude that 'While one may ... be attracted by the fascinating hypothesis ... which cuts the Gordian knot of so many difficult geological problems, ... [it] fail[s] to interpret convincingly the known geological facts'.[11] Two members of the professoriate did not share these attitudes of indifference or antipathy, namely Carey and Jaeger.

Carey studied geology at the University of Sydney in the 1930s. He recalls that David, Cotton and several of David's students and other associates were inclined toward Drift and that he received most of his stimulation from Cotton and David.[12] In 1937, after several years of fieldwork in New Guinea and Papua, he drafted a thesis for the D.Sc. in which he developed his orocline concept[13] and used it to reconstruct Pangaea. Drift was suppressed in the final version for fear of negative responses from overseas examiners.[14] He continued, however, to defend Drift. For example, though Carey and Browne were good friends, they disagreed sharply on this: Carey later called Browne a 'rooted continentalist', while Browne replied by dubbing Carey an 'uprooted continentalist'.[15] After more field-work and war service, Carey was appointed Chief Government Geologist for Tasmania and, in 1946, foundation professor of geology at the University of Tasmania.

Drift featured in Carey's charismatic lectures in the 1940s and 1950s, especially in his course on global tectonics.[16] He warned his students that he would be presenting unconventional views but that this might give them a head start of some twenty years on students elsewhere.[17] This was in many ways not an idle boast, for Carey from the beginning taught a version of Drift very similar to the later seafloor-spreading theory in which the continents were embedded in larger pieces of crust carried along by convection currents in the mantle.[18] He brooded in the 1930s and 1940s on the problems confronting the early versions of Drift, particularly the issues of the motive force and of the 'fit' of the continents, but forbore publication on these matters until he had solved them to his satisfaction. However, his views on Drift were well known: in September 1950, for example, he gave a spirited defence of Drift at a seminar at the ANU. In 1954 he submitted to the *Journal of the Geological Society of Australia* a paper embodying the material omitted from his thesis: a discussion of Drift, his orocline concept and geological reconstructions using oroclines. It was rejected. One of the referees was A.A. Öpik of the Bureau of Mineral Resources (BMR). Although Öpik was a refugee from Estonia, his opinion of Carey's work was probably shared by others in Australia at the time.[19]

S.W.C. completely disregards the geological side ... For an uninformed reader the paper is 'brilliant'. To a person with some knowledge it is grandiloquent obscure ignorance and contempt of past authors. S.W.C. treats his audience of geologists as robots ... [who] will gather the facts and perform the tests. And the master again will pick out the suitable and ignore the rest.

Carey was purportedly so incensed at what he regarded as the high-handed treatment afforded his paper – which he attributed to a deep-seated hostility toward Drift[20] – that he thereafter for many years refused to have anything to do with the national Society.[21] Internationally, the paper came to enjoy some favour after Carey arranged to have it published by the Royal Society of Tasmania.[22] It was to spur Edward

Bullard to produce his famous computer-based 'Bullard Fit' and in 1963 was awarded the Gondwanaland Gold Medal. The term 'orocline' itself was incorporated into standard dictionaries of geophysics.

Carey's most public endeavour on behalf of Drift was the 1956 symposium he convened at Hobart. Voisey and Öpik were explicitly invited and acted as critics of Drift.[23] With the exception of Jaeger, the rest of the Australian earth science professoriate did not accept Carey's invitation to attend. Not surprisingly, the contingent of mainland geologists who did participate (e.g., Voisey, Öpik, J.M. Dickins and G.A. Thomas of the BMR) did not think that the evidence, or more particularly the Australian evidence, supported Drift.[24] In contrast, with the exception of C.A. Longwell from Yale who, though not a Drifter, had long defended Drift as a respectable global theory, the international contingent was made up of Drifters. Ken Caster from the University of Cincinnatti, for example, was one of the few outspoken American proponents of Drift. We might speculate that his presence in Tasmania on an extended study leave, like the organization of the symposium itself, reflected a judgment by Carey that it might be more fruitful to develop international interactions with Drifters than to engage with the mainland community.

The symposium and the subsequent publication of a collection of papers from it[25] did much to launch Carey on the international scene.[26] His own paper, which occupies half the volume, reflected a development of his ideas subsequent to the symposium that led him to abandon Drift and to advocate Expansionism as affording solutions to problems he had encountered in Drift. Carey's view that the major tectonic features of the earth were best explained in terms of an earth that was increasing in size no doubt confirmed his Australian critics in their judgment of his ideas as eccentric. Internationally, however, in North America, Europe and elsewhere, Expansionism had a surge of popularity in the late 1950s and early 1960s and Carey was and remains its best-known proponent in person and in his publications. Even today, Expansionism may be the only serious rival to the plate tectonics version of Drift. In the 1950s, however, in an Australian context, Carey was very much on the periphery, both geographically and theoretically, of the Australian geological community. Carey's only significant ally north of Bass Strait in Drift – and the only group with which he seems to have interacted extensively – was Jaeger's ANU geophysics unit.

Mark Oliphant, the first director of the Research School of Physical Sciences (RSPS) at the ANU, was instrumental in the formation of a geophysics unit and the establishment of a chair in that area. His decision was shaped in part by discussions with J. Tuzo Wilson and Edward Bullard. Oliphant had first learned of Drift – and been inclined to it – at the University of Adelaide in the 1920s.[27] He had subsequently pursued physics at Cambridge and Birmingham. At Cambridge, both he and Bullard had worked in the Cavendish Laboratory but Bullard had subsequently turned to geophysics. Oliphant first met Wilson at Birmingham when the latter gave a paper explaining the formation of island arcs. Oliphant was very favourably impressed by Wilson's stress on geophysics rather than geology as the key to understanding the Earth. Oliphant, after his designation as Director of the RSPS, sought advice from both Wilson and Bullard with respect to the inclusion of geophysics in the School.[28] Wilson, then professor of geophysics at Toronto, featured in a two-day seminar at the ANU in September 1950, soon after Oliphant's arrival and at Oliphant's invitation. Prider, Bryan, Bullen, Carey and others participated. Just over a month later, Wilson supplied Oliphant with a short paper that was to constitute the master plan for geophysics in the RSPS for the first decade of its existence. Wilson began by outlining a distinction between geological mapping or applied geology and 'the broader subject [which] is the study of the earth, how it was formed, what forces operate in it to-day'. It was this latter approach (that of the geophysicist) that he urged be followed at the ANU, expressing the hope that geology soon 'will be on the way towards being a physical science like chemistry or astronomy'.[29] Oliphant apparently took this advice to heart: he later resisted Jaeger's proposals to develop a larger unit embracing applied and exploration geophysics or applied geology.[30] To carry out this task Wilson suggested the formation of 'a team of physicists, geologists and applied mathematicians'.[31] Finally, he counselled the adoption of a laboratory-based programme to pursue the following studies:[32]

Properties of matter at high temperature and pressure.
Temperature and pressure of deposition of minerals.
Radiometric age determinations.
Variation of isotope ratios in different rocks and minerals.
Geothermometry (especially conductivity of rocks).
Collection of data on physical properties of rocks.

Of these Wilson made a special plea for the use of a mass spectrometer in developing new methods (including K^{40}-Ar^{40}) for rock dating. If palaeomagnetic studies fall under the heading of 'physical properties of rocks', Wilson's list charted the research fields that were taken up in the

RSPS, though initially split between the Department of Geophysics and a separate radiochemistry unit. Jaeger, though he did not feel absolutely constrained by Wilson's report, nonetheless acknowledged a few years after his appointment that the scope of the Department had been initially defined by Wilson's report and that he had sought to adhere to its spirit.[33]

Jaeger was not the first choice for the chair and his appointment in 1951 met with criticism. Bullen, for example, who had earlier declined the offer to move to Canberra, thought Jaeger inexperienced in geology and geophysics.[34] Jaeger had won distinction as an undergraduate in mathematics and physics at Sydney and then, like most promising Australian science graduates, had gone to Cambridge for advanced studies. There he had worked in quantum mechanics. After failing to secure a research fellowship at Trinity College he moved physically as far from Cambridge as possible, securing a post in mathematics at the University of Tasmania. He also gave up the 'new physics' and made himself master of the more classical field of heat conduction.[35] He had no formal training in geology or geophysics and in this respect his appointment at the ANU may well have seemed odd. However, he was well known to Oliphant both personally and 'for his work on heat conduction and rock-fracture for the Snowy Mountains Authority'.[36] Oliphant's choice fitted well with Wilson's plan: Jaeger was oriented to physics rather than traditional geology, he was a distinguished applied mathematician and physicist (two of the kinds of expertise needed on Wilson's 'team'), and he had a special interest and good credentials in the field of heat conduction in solids that was readily applicable to geothermometry.

Before leaving Hobart for Canberra, Jaeger decided to acquaint himself with geology and geophysics. Naturally, he attended some of Carey's lectures. This was the first in a series of interconnections between Carey and Jaeger's group at the ANU that revolved around Drift. Although Jaeger was not in these early years a fervent apostle for Drift, he did have 'a very open mind on the whole business', sponsored research in his unit related to Drift and perhaps took a 'puckish delight' when the results 'not exactly antagonized but certainly eventually showed that the continents drifted and a lot of people in the geological establishment in Australia had in fact been wrong'.[37] Jaeger's first appointments were Mervyn Paterson, an engineer and Germaine Joplin, a petrologist. His next, Edward (Ted) Irving, pushed the ANU into the controversy over Drift.

Jaeger had a considerable degree of autonomy and also relatively generous funding at his disposal to implement Wilson's plan. This was related to the charter of the ANU, which was conceived of as the 'pure research' counterpart of the Commonwealth Scientific and Industrial Research Organization. The emphasis was upon international competitiveness in selected disciplines and specialties, notably in undeveloped or newly developing areas. To promote these aims, a generous allowance was made for technical assistance, international travel and visitors. Staff did not have teaching responsibilities. Funding for the ANU came directly through the Prime Minister's office: there were no grants agencies to satisfy, no elaborate review process, no other universities in direct competition for support. This situation contrasted markedly with that of the other Australian universities, which were dependent upon individual state governments for support. That support, especially for research, was meagre. It is no mystery why there would be (and still are!) suspicion, jealousy and rumblings of discontent in the Australian academic community about the privileged status of the ANU. Irving himself remarked on this tension and suggested[38] that it was not eased by the fact that 'the ANU, which was supposed to be the Australian *National* University was really the Australian *International* University. Most of the people who came there [like Irving himself} ... were immigrants from all over the world'.[39] The ANU group did, however, have good working relationships with some staff in the Bureau of Mineral Resources in Canberra, which formed one of the largest groups of field geologists in Australia.

Irving became acquainted with Drift during his schooldays in Britain and went to study geology at Cambridge. While there he worked as a 'tame geologist' for S.K. Runcorn, collecting rock samples for palaeomagnetic measurements. He undertook postgraduate studies with Runcorn and for his thesis topic investigated the palaeomagnetism of the Torridonian sandstones of Scotland.[40] Runcorn prior to 1956 preferred to explain discrepancies between the present pole position relative to Britain and those of the past determined by palaeomagnetic observations in terms of the movement of the magnetic poles. Irving, however, was intrigued by the alternative possibility that the crust had moved relative to the poles, that is, Drift.[41] It occurred to him that measurements of rocks from India, which according to Wegener's theory had drifted thousands of kilometres to the north across the equator after the break-up of Pangaea, might test this possibility. He managed to obtain a few samples and the results were suggestive, though too meagre to publish.

Early in 1954, as his postgraduate work was drawing to a close, Irving applied for a research fellowship at the ANU. The attraction of research in Australia was outlined in his description of his proposed project: it would provide a 'wonderful opportunity to carry out a lengthy experiment' to determine if Australia had drifted.[42] Indeed, his enthusiasm was soon further whetted. K.M. Creer in Runcorn's group had devised a graphical method of representing the motion of the magnetic poles relative to Britain: the first polar-wandering curve. If the curve to be plotted for Australia matched that for Britain, then Runcorn's view would be sustained; if it did not, this would be striking evidence for a relative motion between Europe and Australia. Shortly before Irving was to depart for Canberra, Jaeger visited Cambridge. He and Irving discussed the equipping of the palaeomagnetics laboratory and Irving prepared designs for an igneous astatic magnetometer. Irving was probably more excited about what he heard in a talk Jaeger gave during his visit. Jaeger described preliminary observations he had made on the magnetism of some Tasmanian dolerites dated as Jurassic.[43] Their magnetism was nearly perpendicular to the strata. Irving saw the significance: this indicated that in the Jurassic, Tasmania must have been very close to the south magnetic pole. Therefore, there had been relative movement between Tasmania and that pole. Tasmania was now nearly antipodal to Britain yet, according to its polar-wandering curve, Britain in the Jurassic had been near the equator rather than near the north pole. Therefore, there was a *prima facie* case for substantial relative motion between Tasmania and Britain since the Jurassic.[44] Irving thus had good reason to expect that his proposed studies in Australia could yield results of major importance to the fixist-mobilist debate. He rushed through his thesis and left for the ANU.

Irving, soon after his arrival, asked Jaeger why a department of geophysics was staffed in such a bizarre manner. Why, for example, was there no seismologist? Jaeger replied in the following vein:[45]

It's crazy to get into the classical areas of geophysics [seismology, gravity, geomagnetism] because you need enormous amounts of money and you need at least ten years before you are in the least competitive. But what do we want to know about the earth? We want to know something about its behaviour at depth, so we have to build apparatus to look at high pressures. You don't hire a geophysicist to do that, you hire an engineer to build the apparatus [Paterson]. You have to talk to geologists so you have to know what the names of rocks are [Joplin]. In the long term the earth is a heat engine and perhaps I [Jaeger] might be able to make a contribution and there are not too many people doing it. And you [Irving], you're in a completely new field – you might even turn up something interesting. And what's more, if you don't, I can fire you after three years. On the other hand, what you do may turn out to be rather important and even so it won't cost me very much.

Although palaeomagnetism had not been listed explicitly in the 'Wilson plan', this was certainly an interesting 'physical property of rocks' that might have considerable significance. Jaeger himself had probably been alerted to the larger issues that could be addressed – for example, deciding between P.M.S. Blackett's 'fundamental' theory of geomagnetism and the 'internal dynamo' theory advocated by Bullard and others – during a visit by Blackett to Australia in late 1953 and early 1954. Jaeger arranged through Blackett to have some dolerite samples processed in the latter's laboratory.[46] Blackett himself had done pioneering work in directionalist palaeomagnetic studies and favoured Drift, in part perhaps because of the case made by Carey when he visited Tasmania. Jaeger's investment in palaeomagnetism was to pay good dividends.

Irving visited Tasmania early in 1955 to collect samples of dolerite and there met Carey, who had recently returned from a study leave at the ANU. Carey briefed him on the geology of Tasmania and helped him choose sampling locations. Irving's first results were in agreement with Jaeger's. He swiftly prepared two papers and submitted them to the *Journal of the Geological Society of Australia*. One was based on the dolerite work. He argued that the Jurassic position of Tasmania lay near the South Pole, that this was supported by palaeoclimatic evidence, that it did not match up with the polar wander curve for Europe and that this could be explained by significant relative movement between Tasmania and Britain. He pointed out that this did not constitute proof of Drift: his conclusions embodied a number of assumptions and rested on a still small body of data. Nonetheless, he opined that palaeomagnetic measurements in Australia 'ought to provide a critical test for the hypothesis of continental drift.[47] This was quite a large bite to be swallowed. Irving had just landed in Australia yet was already offering a sweeping reinterpretation of Australia's geological history. He was associated with an institution that was not very popular within the Australian academic community. He was employing techniques and arguments that were both novel and controversial. Finally, his apparently favourable attitude towards Drift ran counter to the views of many members of the Australian geological establishment. The dolerite paper was rejected by the *JGSA* referees. Through the good offices of Carey, it was published instead by the Royal Society of Tasmania.

Irving commented that Jaeger had been enormously helpful in preparing the second paper, which was based on Irving's thesis and which discussed in general terms palaeomagnetic directionalist studies, Drift and palaeoclimatology. He showed a draft to Jaeger who 'took me by the arm, escorted me to his office, sat me down, and proceeded to give me a two-hour session on how to write papers "because I don't want to have to bloody well tell you every time"'.[48] Alderman and Browne[49] refereed the paper for the *JGSA*. This one too was rejected. Irving then submitted it to the international journal *Geofisica pura e applicata*, which published it in 1956.[50] In it, Irving clearly and concisely spelled out the major contentions of and the research agenda for palaeomagnetic investigations relating to Drift that were to be pursued by researchers in Britain and elsewhere over the next few years. In particular, he argued that palaeoclimatic evidence could and should be used to verify palaeomagnetic evidence.[51] Secondly, he contended that pole paths for different continents were different and therefore there was not only relative motion between the poles and the continents but also relative motion among the continents.[52]

During the rest of his stay in Australia, Irving did much to carry out the agenda he had set. His presentation to other Drifters at Carey's 1956 symposium seems to have been well received, although Irving tried to dissuade Carey from terming the palaeomagnetic evidence 'proof' of Drift.[53] Neither Jaeger nor Irving subsequently followed Carey down the path of Expansionism. After Irving's rebuffs by the *JGSA*, he published most of his papers in overseas journals used by other palaeomagneticians: *Nature, Geophysical Journal of the Royal Astronomical Society, Journal for Geophysical Research, Advances in Physics* and so on. His work survived the scrutiny of Allan Cox and Richard Doell in their influential and sceptical review of the palaeomagnetic literature and indeed they pointed to the Australian data as 'evidence for a relative displacement of Australia with respect to North America and Europe which cannot be ignored'.[54] The development of sophisticated rock-dating facilities in the Department of Geophysics enabled Irving to introduce a greater degree of precision and accuracy in his own work. Instead of using samples of igneous rock for which the dates were often dubious, he was able to base his polar-wandering curves on K-Ar dated outcrops. Indeed, in a series of papers published by his group in 1963, 'in many cases K-Ar determinations were made on the *same* rock sample ... studied palaeomagnetically.[55] His conclusions were, however, unaltered: 'the only reasonable explanation for these observations is either that the earth's field during the Mesozoic was nondipolar or that Australia has moved relative to these northern continents by approximately 3000 to 5000 km. during the past 100 m.y.'.[56] These papers made an impression even on some of those previously sceptical of Drift.[57] John Verhoogen at Berkeley, for example, noted that the Australian results had been among the most compelling evidence for Drift.[58]

Another area of palaeomagnetic research was pursued by Frank Stacey. Stacey, a Londoner, had completed a PhD in 1953 at the University of London on the effects of very high pressure on ferromagnetism and had been a research fellow at the University of British Columbia. He had had no interest in geophysics until 1956. At about the same time that he responded to an advertisement for a research fellowship in Jaeger's unit, he read an article by Louis Néel[59] on the theory of rock magnetism. In his application to Jaeger, he proposed to pursue the implications of Néel's ideas and to develop a sophisticated theory of thermo-remanence.[60] Jaeger must have been delighted: Stacey had interests in two of the areas advertised, high pressure physics and the physics of magnetism. His appointment fitted not only with the Wilson plan in respect of the collection of data on the physical properties of rocks and of the effects of high pressures, it was also to prove complementary to Irving's interests.

In the mid- and late 1950s, John Graham at the Carnegie Institution of Washington's Department of Terrestrial Magnetism challenged a central assumption of palaeomagnetic researchers. Graham, who was soon joined by A.F. Buddington of Princeton and J.R. Balsley of the US Geological Survey, contended that remanent magnetism acquired by rocks during directed stress might be significantly altered when that stress was removed, for example when a sample was taken for measurement in a magnetometer. Were this the case, then there might be no correlation between palaeomagnetism as measured in the laboratory and the palaeomagnetic field at the time the rock was formed. This effect of magnetostriction would mean 'that many conclusions that have been offered on the basis of rock-magnetism data, relating to polar wandering, continental drift, secular variation, and reversal of the earth's magnetic field, are subject to serious doubt'.[61]

Stacey, soon after his arrival in Canberra, read Graham, Buddington and Balsley's 1957 paper on magnetostriction. This he assessed to be a more serious threat to palaeomagnetic studies than reversals or Néel's theoretical arguments.[62] Irving's reaction was that magnetostriction could not be a general effect: there was too much consistency among measurements on rocks from different locations, with different composition, with different kinds of stress, and so on.[63] Stacey, together with Peter Stott, a postgraduate student from England working under Irving, carried out a series of experiments on the effects of stress on palaeomagnetism. From this work issued a series of papers rebutting Graham's attack.[64] These papers, published in *Nature* and the *Journal of Geophysical Research*, seem to have been well received by most American and British palaeomagnetists – nor, perhaps, should that be surprising since it overcame one source of objections to the validity of their specialty. Cox and Doell, for example, in their review, seem to have regarded the issue of magnetostriction as closed.[65] Both Stacey and Irving soon departed the ANU, Stacey in 1961 to take up a position in geomagnetism with the Meteorology Office in Cambridge, Irving in 1964 for Canada.

Rock dating using K-Ar and other isotopic methods was a second research area at the ANU that was related to Drift, though initially less directly than palaeomagnetic polar determinations.[66] The 'Wilson plan' had given prominence to rock dating using isotopic methods and a unit that included John Richards had been established in the RSPS at about the same time as the Department of Geophysics. By the late 1950s, however, there was some dissatisfaction with the unit: it had not produced copious publications or significant breakthroughs. The techniques employed (Pb-isotope dating using galenas; zircon dating by the alpha-counting method) were by the late 1950s crude and outmoded. In contrast, K-Ar methods and equipment developed at Berkeley were revolutionizing Pleistocene chronology and were also applicable to earlier geological periods. Many professional and personal ties were established from the early 1950s between researchers connected with Berkeley and those at the ANU. These ties were centred upon techniques and data rather than upon such global theories as Drift. Francis Turner visited the ANU and met Paterson who, on Turner's recommendation, went to Stanford to work with David Griggs on rock deformation. In 1958 Garniss Curtis, one of the Berkeley rock-dating group of Jack Evernden, Curtis and John Reynolds, visited the ANU and went sample-collecting with Irving. While there, he gave a talk on K-Ar rock-dating. Jaeger was fired with enthusiasm and proposed to Curtis that he return to Canberra to set up a K-Ar laboratory and that someone from the ANU go to Berkeley to learn how to run this.[67] In 1960 Jaeger visited Berkeley. In the meantime, the radiochemistry unit had been broken up, and Jaeger had acquired W. Compston, Richards and 'a pot of money' that facilitated a move into K-Ar dating. He had also secured agreement from the BMR to provide some technical assistance. With the ground thus prepared, Jaeger and the Berkeley group put in train his earlier proposal. Evernden agreed to go to Canberra for several months to set up the K-Ar laboratory. The entire apparatus including the Reynolds mass spectrometer and the argon extraction line was constructed in Berkeley and shipped to Canberra for assembly. Thus, the laboratory would be a duplicate of the one at Berkeley. Evernden while in Canberra would also train Richards. Finally, Ian McDougall, who was finishing a thesis in petrology at the ANU, would be sent to Berkeley for training so that he could on his return operate the new laboratory and direct research in this area.[68]

Evernden was eager to go to Australia: outcrops of good fossil-controlled datable rocks from the Mesozoic and Palaeozoic were rare in North America but there were excellent horizons in Australia, especially in Victoria.[69] He quickly set up the laboratory, a process that mostly involved unpacking and assembling the components shipped from Berkeley. This left him time to go sample-collecting with Irving. Evernden became aware of the resentment of state universities towards the ANU. He later described how he and Irving 'sneaked across the border' into Victoria for about two weeks, collecting good datable palaeozoics without contacting Victorian geologists: 'Those guys didn't know rocks had been dated until John [Richards] and I gave a paper on it at the ANU near the end of my stay'.[70] This paper[71] proposed a wholesale revision of the former history of eastern Australia based on K-Ar dating rather than dating on the basis of 'rock-type' that Browne had sometimes used.[72] Many of the dates given for igneous rocks in David's *Geology of Australia* differed significantly from the K-Ar dates. In fifty-five igneous intrusions examined, Evernden and Richards found a difference of a geological *period* or more in forty-one cases.[73] Their results, which indicated nearly continuous volcanic activity during the Palaeozoic and Triassic, meant that Australia could have major importance as an exporter of K-Ar dates: 'it is to be expected that the ultimate contribution of Australia to the establishment of the geological time-scale in the time range 200–600 m.y. will be larger than that of any other continent'.[74]

McDougall for his part was eager to go to Berkeley. When Jaeger proposed to him that he take up K-Ar dating, his career was nearing a fork: he had almost finished his thesis on the petrology of the Tasmanian dolerites and had to decide what to do next. He might seek a position at the ANU or do postdoctoral studies in petrology in Britain. Now he was offered not only the opportunity of acquiring a new set of techniques but also the opportunity of employment. Jaeger indicated that although an additional appointment in petrology was unlikely, there would be one in rock-dating. Jaeger's 1958 proposal and his orchestrations to implement it soon proved successful. McDougall left for Berkeley on a CSIRO fellowship. He arrived soon after Evernden had left for Canberra. Coincidentally, he began his K-Ar training under Reynolds on the same day as Brent Dalrymple, who was soon to become the 'dater' for Cox and Doell. These were to be McDougall's chief competitors in the production of polarity-reversal timescales. He met Cox and Doell at the US Geological Survey at Menlo Park near Berkeley and had a vague discussion with them about geomagnetic field reversals and establishing a time scale based on them; but this for McDougall lay dormant until some time after his return to the ANU.[75] More memorable for him was a discussion at Princeton in early 1961 with the geophysicist Harry Hess in which Hess sketched out his new version of Drift based upon seafloor spreading.[76] McDougall quickly came to share the enthusiasm of his Berkeley colleagues for K-Ar dating techniques but it was otherwise with respect to their global theories.

McDougall, prior to his visit to Berkeley, had been steeped in Drift. He had completed an honours degree in geology at the University of Tasmania where he had adopted Drift as a result of Carey's lectures.[77]

> I don't think there was any question about not accepting this hypothesis; that the data – particularly the way Carey presented it – were almost incontrovertible. Carey had such a happy way of putting it all together that it was extremely persuasive, and I don't think there was ever a stage at least in my formative years of having much in the way of doubts about it, and why the rest of the world wasn't with it was very difficult to see from that perspective.

In his final year in Hobart he had attended Carey's symposium, at which he was introduced by Carey to Jaeger. We might speculate that Carey had already decided on the ANU for McDougall. McDougall recalls that Carey insisted that all his students go elsewhere for their higher degrees. The usual practice in most Australian geology departments at that time was that if one's students were really bright, they should be sent off to Cambridge; if they were only 'pretty good', they could stay on in their department; but under no circumstances should they be sent to the ANU.[78] Carey, however, encouraged McDougall to go to the ANU, where he chose a thesis topic that drew together his interests in petrology and Drift: a petrological study of the Tasmanian dolerites, the formation of which he thought might be connected to the break-up of Pangaea.[79] His sympathy for Drift, stimulated by Carey and sustained at the ANU, was not shared at Berkeley. While at Berkeley, he gave a talk on Drift to the graduate students. Their response was incredulity; they regarded Drift 'as all a bit of a joke', claimed that there was little basis in fact for these ideas, and so on. McDougall recalls his reaction to this as follows:

> I was just amazed; I couldn't believe it ... [the evidence was persuasive when Drift was proposed] and by Carey's time it was even more persuasive, and it seemed to me that any rational-thinking individual could not dismiss this remarkable amount of consistent data out of hand and yet it was.

This negative attitude also surfaced when he referred to Drift in a talk to the Berkeley department on the Tasmanian dolerites. His work in K–Ar dating was not to be met with the same incredulity.

After his return to the ANU in 1961, McDougall was initially fully occupied with dating dolerites and with meeting the demand from Irving and others within his unit for K-Ar dates. However, a few months before the publication in June 1963 by Cox, Doell and Dalrymple of their first reversal timescale, McDougall began collaborative work on this topic with Don Tarling, an English palaeomagnetist interested in Drift and a PhD student of Irving. Their first timescale appeared in *Nature* only a few months after the first scale of the Menlo Park group.[80] A friendly competition ensued and over the next few years the researchers at the ANU and Menlo Park leap-frogged one another in producing ever more detailed and extended timescales in *Nature*, the *Bulletin of the Geological Society of America*, and other 'international' publications.[81] It is notable that in this competition, the only dates other than their own that the Menlo Park Group took at face value in constructing their successive scales where those of McDougall, and *vice versa*.[82] By the mid-1960s the ANU had established itself as one of the major international centres in this field.

The familiar dichotomies of centre-periphery, metropolitan-colonial, national-international or dependence-independence used to describe science in Australia and its relationship to science elsewhere do not seem to apply readily to the developments outlined above. Perhaps this is owing to major changes after the Second World War in the character of scientific institutions and practice in Australia that invalidate comparisons with earlier eras. The ANU, for example, was intended to be and was a special institution with special aims in the Australian context: as Stacey put it, 'the ANU was set up essentially to leaven the lump'.[83] However, as a matter of procedure I prefer not to reach for some magic sword inscribed 'colonial science' to cleave through the tangled skein of cognitive and social threads that make up so-called 'Australian science'. Instead, I prefer to seek first to analyse instances of scientific practice there, using the same sorts of tools that I might apply to science elsewhere: to name a few, issues of theory change and theory choice, specialization, research networks and the ubiquity of cognitive and social interests. Any work then to be done by those grand dichotomies – and the sweeping and sometimes unexamined generalizations implicit or explicit in them[84] – comes as the result of an analysis, not as its premise. With respect to the research pursued by Carey, Irving, Stacey, Jaeger and McDougall and the reception accorded their work by others, it seems more helpful to consider the research networks in which they participated and the processes by which their knowledge claims were judged within and beyond those networks.

The research networks of Carey and the ANU researchers were, depending on the problem-fields involved, made up of shifting memberships drawn in varying proportions from Australian and other institutions. Stacey, for example, except when he engaged in some co-operative research efforts with Irving or John Lovering, had almost no interaction with other scientists in Australia and certainly no interaction with Australian geologists. Carey added to his ties with the ANU relationships with Drifters in other countries, once he began publicly to espouse Drift through his symposium, his publications, his correspondence, and his travels. McDougall, through his training at Berkeley and his subsequent work at the ANU, became a major participant in the circle of K-Ar daters and in the small group of those involved in the development of reversal timescales. Irving maintained his ties with researchers in Britain after his move to Australia and forged new ones with Australian and North American workers. Jaeger had personal and professional ties with Carey; as head of the geophysics unit he took a close interest in the work of his colleagues; he exploited earlier connections such as those he had formed with Wilson and Bullard; and he established new ones related to the 'Wilson plan' and its implementation such as those with the Berkeley group and with Blackett. These varied networks represented shifting, complex relations of interdependence. For example, Jaeger was initially heavily dependent on overseas links for techniques, instruments and trained personnel. Moreover, both the ANU group and Carey looked overseas rather than to the local geological community for judgment on and recognition of their knowledge claims. Within the Australian community, both the ANU group and Carey possessed a considerable measure of institutional independence, which enabled them to pursue preferred lines of research and to promote Drift interpretations that ran counter to the views of most Australian geologists. This is especially marked in the institutional structure and funding of the ANU which made it possible, for example, for Jaeger to acquire the expertise on which his research programme depended. It could also be argued that the ANU workers and Carey were aided in maintaining their independence by the recognition that their work received from the international networks in which they participated.

Overseas participants in these networks were in turn dependent on Australian members in several respects. First, they relied chiefly upon Australian workers for data on this geologically important region. These data – Irving's polar-wander paths, Stacey's experimental rebuttal of the magnetostriction argument, points on the reversal timescale generated by McDougall – were disseminated both informally through correspondence, delivered papers and face-to-face discussions, and formally through international publications. Secondly, overseas participants received some exports of a more theoretical kind: the orocline concept, Carey's version of Expansionism, Irving's linking of palaeoclimatology with palaeomagnetism and the linking of both with Drift. Thirdly, the ANU provided a base for graduate and postdoctoral students from Australia and overseas, for example McDougall, Stott and Tarling; and for visiting researchers such as Turner, Curtis and Evernden: both of these functions are likely to have further strengthened the meshes of the various networks. Fourthly, overseas members of these networks were as dependent upon Australian workers for recognition of their knowledge claims as the latter were upon the former for recognition of theirs: Cox, Doell and Dalrymple upon McDougall; Graham, Buddington, Balsley and others upon Stacey; British constructors of polar-wander paths upon Irving. Although I have described these relationships within the various research networks in terms of *dependence*, a more accurate characterization would be *mutual dependence* or *interdependence*. Although the relations of 'strength' between, say, Irving and the rest of the polar-wander network (Runcorn, Blackett, Clegg et al.) might be unequal in terms of recognition, authority, power and so on, nonetheless there were elements of interdependence:

after all, no scientist (or knowledge claim) is an island unto himself (herself, itself), even though that scientist might be on the 'world's largest island' (Australia) or a much smaller one (Tasmania).

The acceptance of the knowledge claims put forward by Australian researchers is closely related to their participation in these international research networks. Once grafted into the ANU Department of Geophysics, polar-wandering and rock-dating work was almost immediately on a par with – and seen by international workers to be on a par with – work done at Berkeley, Cambridge and elsewhere. This should not cause surprise. The researchers at the ANU, the research programmes they pursued, the equipment and techniques they used, and the theories they employed were familiar to the overseas participants in these networks. As we have noted, the research agendas of Stacey, Irving and McDougall – and perhaps we could extend this via the 'Wilson plan' to the ANU department as a whole – were set 'overseas' and developed and pursued in an international context. Jaeger, in order to make the Department quickly competitive as the Australian research centre in geophysics, and presumably with Oliphant's backing, did not initially concentrate on nurturing local talent. Instead, he simply purchased from overseas the expertise and equipment needed. Both the people and the machines were familiar to workers elsewhere. The favourable overseas judgment of Irving's work was surely related to the fact that he was trained in and well-known to the British community of directionalists, and, for the American audience for his 1963 papers, to the fact that the dates for points on his polar-wander curves for Australia were supplied in part by the Evernden-Curtis team at Berkeley and in part by McDougall and Richards using a 'Berkeley' laboratory. Stacey had been trained – as a physicist – in London and was responding directly to a challenge to the very foundation of palaeomagnetic studies. McDougall had been trained by the Berkeley group and used Berkeley instruments and techniques in his rock-dating work and in his competition over the reversal timescale. Moreover, he had established personal contacts with Reynolds, Evernden, Curtis, Dalrymple and other overseas workers in the field. Finally, Irving, Stacey and McDougall published their key papers in the same journals used by other members of their respective networks and they cited the appropriate literature. As for Carey, the volume based on his 1956 symposium was well received by those who were not antipathetic toward Drift; his defence of Earth expansion gained considerable notoriety but was eclipsed in the late 1960s and early 1970s by the plate tectonics version of Drift.[85] We might conjecture that Carey's publications, though extensive, were perhaps less instrumental in disseminating his ideas than his lecture tours and personal contacts around the world.

Were there any distinctively 'Australian' elements for which we must find a place in this form of analysis? Two come to mind. First, the Australian community of academic earth scientists was numerically rather small compared with those of North America and Britain and, moreover, it was relatively more dispersed: in the 1950s, the pattern was roughly one university per state and most of these were separated from one another by hundreds if not thousands of kilometres. In so far as researchers at the ANU were pursuing relatively novel problems using relatively new techniques, there were few if any fellow specialists in Australia with whom to interact. It was as easy, and perhaps far more productive, to participate in international rather than Australian research networks. As Stacey put it, 'I was interacting with other geophysicists in other places but in Australia there hardly were any'.[86] Perhaps the same could be said of Carey with respect to theoretical issues: except for the ANU group, there were few Drifters in Australia. Secondly, the very location of these Australian researchers gave them an advantage in participating in international research networks and perhaps a special weight to their arguments. Australia in this period and afterwards was of special interest to researchers elsewhere because of its indigenous raw materials, that is, its special geological features that made it a valued source of data *re* Drift and *re* geochronology. Australia's rocks and geological history were in the 1950s and early 1960s as intriguing as its bizarre flora and fauna had been in the nineteenth century. It was this 'colonial' aspect that attracted interest, visitors and emigrants from North America and Britain: Curtis, Evernden, Cox, Doell, Dalrymple, Caster, Irving, Tarling, and many more. Jaeger himself appreciated the advantages Australia offered in competing with northern hemisphere scientists. In replying to Irving's query about research directions at the ANU, he commented:[87]

> If you want to do anything original, you have to go where people aren't. You don't compete. You look at the Magellanic Clouds or kangaroos or something in Australia, you don't go looking at polar bears or the North Star. You look at those things where your position gives you a natural advantage.

We might conjecture that this 'natural advantage' gave the pronouncements of researchers in Australia greater weight in some circumstances. For example, although Irving was but one member of a network of twenty or so researchers engaged in polar-wander studies, he was the only 'Australian' representative. As such, he was in a sense speaking for an entire continent and, as such, he could truly speak univocally on the presumed drift of that continent or on general matters involving a presumed motion of it. There was no need to seek consensus with other experts in Australian polar wandering for there were none.

I close by returning to the common wisdom with which we began. If it be ill-founded, how did it come to be a commonplace?[88] What purpose does it serve? There is some basis for it: in Australia, one could point to Carey and researchers at the ANU; in South Africa, to the defence of Drift by Alexander du Toit and later by Lester C. King, and to the research in the 1950s of the palaeomagnetics group at the Bernard Price Institute. Perhaps their work prior to the 'revolution' was both unorthodox and controversial relative to that of their colleagues and therefore more memorable; perhaps after the 'revolution' their work took on significance it had not previously had, at least among former non-believers. The possible basis for the commonplace is, however, of less interest than the purpose it may serve. It provides a satisfying account of why the majority opinion in the international community from the 1920s through the mid-1960s was, as judged from today, wrong; for it shows it to have been wrong for good reasons. That is, it suggests that the empirical evidence for Drift was far stronger in the southern hemisphere or in Australia than elsewhere but since that evidence was not well-known to the international community, it was not taken into account in reaching a negative judgment of Drift. In this account, earth scientists in the south had the evidence under their noses, so to speak, and therefore reached what would now be regarded as the 'correct' judgment – which would have been reached earlier by northern hemisphere geologists had they gone to live and work 'down under'. This invokes further premises that are also questionable. It is enough perhaps to say that the commonplace and the assumptions that underlie it well serve one purpose: a mythical view of both science and scientists is neatly preserved by the myth of the southern acceptance of Drift.

Department of History and Philosophy of Science
University of Melbourne
Australia

NOTES

* I gratefully acknowledge comments on an earlier draft received from D.F. Branagan, S.W. Carey, R.W. Home, J. Sapp and T.G. Vallance. I appreciate the co-operation of those many individuals cited in the notes who shared with me their recollections and papers and of W. Glen who permitted use of some interview materials. This research was supported by ARC grant A58716008 and the Research Fund of the Faculty of Arts of the University of Melbourne.

1 H.E. Le Grand, 'Specialties, Problems and Localism: The Reception of Continental Drift in Australia, 1920–1940', *Earth Sciences History*, 5(1986), 84–95; H.E. Le Grand, *Drifting Continents and Shifting Theories* (Cambridge, 1988).

2 Le Grand, op. cit. (n. 1: 1988).

3 Ibid.; H.E. Le Grand 'Paleomagnetism and Continental Drift: Historical Introduction', in *Encyclopedia of Solid Earth Geophysics*, ed. D.E. James (New York: 1989), 901–909; pp. 906–907.

4 See Le Grand, op. cit. (n. 1: 1986).

5 H. Frankel, 'Biogeography, before and after the Rise of Sea Floor Spreading', *Studies in History and Philosophy of Science*, 15 (1984), 141–68; Le Grand, op. cit. (n. 1).

6 Drift continued to be treated more sympathetically at Sydney than at other 'state' universities even in the post-war years. Cotton lectured on Drift into the late 1940s and du Toit and Wegener were read by many of the advanced students (David Branagan, private communication).

7 Frankel, op. cit. (n. 5); Le Grand, op. cit. (n. 1: 1986).

8 Frankel, ibid.

9 E.S. Hills, personal communication (hereafter, p.c.).

10 A.L. Hales, 'Keith Edward Bullen, 1906–1976', *Historical Records of Australian Science*, 4(2) (1979), 43–64.

11 T.W.E. David, *The Geology of the Commonwealth of Australia; edited and much supplemented by W.R. Browne*, vol. 1 (London, 1950), pp. 688–89. T.G. Vallance suggests (p.c.) that the "woolliness" of Drift theory in Browne's day was alien to his systematic, logical style and may have contributed to his distaste for that theory; see also Vallance, 'William Rowan Browne, 1884–1975', *Historical Records of Australian Science*, 4(1) (1978), 65–81.

12 S.W. Carey, *Theories of the Earth and Universe: A History of Dogma in the Earth Sciences* (Stanford: 1988), p. 95.

13 Oroclines were patterns of folding produced by rotations of crustal blocks; for example, the rotation of Spain resulted in folding of the Pyrenees at the "hinge" of the rotation. By straightening oroclinal curves and folds, Carey produced a jigsaw fit superior to Wegener's, without introducing the seemingly arbitrary distortions of modern continental shapes on which Wegener had relied.

14 S.W. Carey, *The Expanding Earth* (New York: 1976), pp. 2–3; Carey, p.c.

15 Carey, p.c.

16 J. Elliston, p.c.; Ian McDougall, interview by H. Le Grand, 25 March 1988, t1 s1.

17 John Elliston, p.c. Elliston was one of Carey's first students, completing his honours degree in 1950.

[18] Elliston, p.c.; Carey, interview by H. Le Grand, 29–31 August, 1988.

[19] Öpik to M.F. Glaessner (editor of *Journal of the Geological Society of Australia)*, 1 November 1954; Öpik papers, Australian Science Archives Project (University of Melbourne). The probability that Carey's advocacy of Drift rather than personal factors resulted in this rejection is strengthened by the fact that the same panel of referees (Alderman, Browne, Hills) had previously approved for publication another paper by Carey ['The Rheid Concept in Geotectonics', *JGSA*, 1 (1954), 67–117]. That earlier paper, though synthetic and theoretical and invoking convection currents in the mantle, had not explicitly mentioned Drift.

[20] In such matters it is difficult to disentangle the personal from the cognitive: Carey's brilliance as a lecturer may have also irritated his less flamboyant colleagues.

[21] Carey, p.c.; McDougall, 25 March 1988, t1 s1.

[22] S.W. Carey, 'The Orocline Concept in Geotectonics', *Papers and Proceedings of the Royal Society of Tasmania*, 89(1955), 255–288. Offprints from this small-circulation journal were circulated among overseas geologists and geophysicists with an interest in Drift.

[23] Öpik papers, op. cit. (n. 19).

[24] A caveat must be issued here. Although it is not unreasonable to presume that some, even many, members of the Australian geological community may have had a low opinion of Carey's work because of their hostility to Drift, it may well have been the case that this was reinforced by a distaste for his style and approach, which differed markedly from the norms of narrow empiricism, very cautious inductivism (i.e., what those with a more global outlook sometimes describe as a 'stamp-collecting' mentality), and deliberate understatement of interpretations. Öpik, for example, who had reacted so sharply to Carey's orocline paper, in the notes for his summation at the symposium decried the palaeoclimatological and other evidence for Drift but nonetheless commented 'I personally think that the horizontal displacement of parts of the crust in the geological past, as seen in the present continents, is probable or even evident, and that the present structure of the earth's surface cannot be explained satisfactorily without consideration of such a displacement, – the Continental Drift' Öpik papers, ibid.).

[25] S.W. Carey, ed., *Continental Drift: A Symposium* (Hobart: 1958).

[26] V. Vacquier at Scripps commented (interview by W. Glen, 23 April 1979, t1 s1) that 'at the time, [Carey's] was the best exposition of the data on continental drift'; similarly, Fred Vine recollects (interview by W. Glen, 18 May 1979, t1 s1) that he 'was very impressed by [Carey's] 1958 paper' and that Carey's ideas were widely discussed at Cambridge. On the other hand, John Belshé, who had visited the ANU, commented in a letter (Scripps Institution of Oceanography, Archives, Biographical Files AC5 Box 1, 'Belshé', letter dated 3 November 1960) that 'Carey was in Cambridge while I was away and left the impression of being a "mad-man"'.

[27] The occasion was a debate between F. Wood-Jones, an anatomist and biologist who stressed the value of Drift in problems of species distribution, and Douglas Mawson, a geologist who dwelled upon the implausibility of the continents moving bodily through the ocean floors (M.L.E. Oliphant, 'Formal Opening', in *The Expanding Earth: A Symposium*, ed. S.W. Carey (Hobart, 1981), pp. vii–viii; also, Oliphant, p.c.).

[28] Oliphant, p.c.
[29] J.T. Wilson, 'Notes on the Study of Geophysics by Professor J. Tuzzo [sic] Wilson. University of Toronto – November 1950'; ANU Archives, Administrative Records (Registrar's Division), document 655/1956, p. 3.
[30] M.S. Paterson, 'John Conrad Jaeger, 1907–1979', *Historical Records of Australian Science*, 5(3) (1982), 65–88; pp. 76–77.
[31] Wilson, op. cit. (n. 29), p. 5.
[32] Ibid., p. 6.
[33] J.C. Jaeger, Letter dated 25 January 1955 to the Vice-Chancellor, ANU; ANU Archives, file 8.7.1.1 (Department of Geophysics Policy).
[34] Oliphant, interview by M.S. Paterson, 22 December 1980. Oliphant (p.c.) himself was not enthusiastic about Bullen on the grounds that his interests in geophysics were circumscribed and that 'he was utterly opposed to the whole concept' of the ANU.
[35] Paterson, op. cit. (n. 30). Jaeger discussed his own failed ambition at Cambridge and his retreat to Tasmania and classical physics when counselling E. Irving after the latter had learned that his PhD thesis had been rejected (E. Irving, reminiscences on Jaeger to M. Paterson, January 1981, t1 s1).
[36] Oliphant, p.c.
[37] Irving, loc. cit. (n. 35).
[38] E. Irving, interview by H.E. Le Grand, 12 February 1988, t1 s1.
[39] This tension, together with a general antipathy toward Drift, may have contributed to negative judgments by Australian earth scientists of some of the work conducted in Jaeger's department.
[40] E. Irving, 'The Palaeomagnetism of the Torridonian Sandstone Series of North-Western Scotland', PhD thesis, University of Cambridge, 1954.
[41] See Le Grand, op. cit.(n. 1: 1988), pp. 138–169, for a discussion of the development of palaeomagnetic studies and their relation to Drift.
[42] Irving, op. cit. (n. 38), t1 s1.
[43] See J.C. Jaeger and G. Joplin, 'Rock Magnetism and the Differentiation of Dolerite Sill', *Journal of the Geological Society of Australia*, 2(1955), 1–19.
[44] Irving, op. cit. (n. 35), t1 s1.
[45] Irving, interview by W. Glen, 7 February 1978, t2 s1; also, Irving, op. cit. (n. 35), t1 s1 and (n. 38), t1 s2.
[46] J.C. Jaeger papers, Basser Library, Australian Academy of Science, MS 120/1.
[47] E. Irving, 'The Magnetization of the Mesozoic Dolerites of Tasmania', *Papers and Proceedings of the Royal Society of Tasmania*, 90(1956), 157–168.
[48] Irving, op. cit. (n. 35), t1 s1.
[49] There was no personal animosity on Browne's part. Even though he was a staunch foe of Drift, he became good friends with Irving and went with him on sample-collecting trips. Irving ascribes Browne's rejection of his 1956 paper in part to his distrust of palaeoclimatology, a distrust that dated from the 1920s and 1930s, and that was founded on Browne's detailed knowledge of inconsistencies and outright misidentifications of deposits in the palaeoclimatic literature on Australia. Irving jokingly remarked that he himself 'had no alternative but to believe in palaeomagnetism – there was nowhere else to go!' (Irving, op. cit. n. 38, t1 s2).
[50] E. Irving, 'Palaeomagnetic and Palaeoclimatological Aspects of Polar Wandering',

Geofisica pura e applicata, 33 (1956), 23–41.
[51] Ibid., pp. 35–7. Irving (ibid., p. 40) credits Jaeger with suggesting this. However, the argument is clearly set out in Irving's thesis (op. cit. [n. 40], p. 110) and he now regards the published note as mistaken (Irving, p.c.).
[52] Irving, op. cit. (n. 50), p. 40.
[53] E. Irving, 'Rock Magnetism: A New Approach to the Problems of Polar Wandering and Continental Drift', in Carey, op. cit. (n. 25), pp. 24–61; p. 5; also, Carey, ibid., p. 357 and Irving, op. cit. (n. 38), t2 s1.
[54] A. Cox and R.R. Doell, 'Review of Paleomagnetism', *Bulletin of the Geological Society of America*, 71 (1960), 645–768; p. 762.
[55] E. Irving, W.A. Robertson and P.M. Stott, 'The Paleomagnetism of some Mesozoic Rocks from Eastern Australia: Preliminary Remarks', *Journal of Geophysical Research*, 68 (1963), 2281–2; p. 2281.
[56] E. Irving, W.A. Robertson and P.M. Stott, 'The Significance of the Paleomagnetic Results from Mesozoic Rocks of Eastern Australia', *Journal of Geophysical Research*, 68 (1963), 2313–7; p. 2313.
[57] Irving, op. cit. (n. 38), t2 s1.
[58] J. Verhoogen, 'North American Paleomagnetism and Geology', in *Geological Society of America Centennial Special Volume*, ed. E.T. Drake and W.M. Jordan (Boulder, 1985), 401–7; p. 403.
[59] L.E.F. Néel, 'Some Theoretical Aspects of Rock Magnetism', *Advances in Physics*, 4 (1955), 191–243.
[60] F.D. Stacey, interview by H.E. Le Grand, 19 May 1988, t1 s1.
[61] J. Graham, A.F. Buddington and J.R. Balsley, 'Stress-induced Magnetizations of Some Rocks with Analyzed Magnetic Minerals', *Journal of Geophysical Research*, 62(1957), 465–74.
[62] Stacey, op. cit. (n. 60), t1 s1.
[63] Ibid.
[64] F.D. Stacey, 'Effect of Stress on the Remanent Magnetism of Magnetite-bearing rocks', *Journal of Geophysical Research*, 63 (1958), 361–8; Stacey, 'Magnetic Anisotropy of Igneous Rocks', 65 (1960), 2429–42; P.M. Stott and F.D. Stacey, 'Magnetostriction and Palaeomagnetism of Igneous Rocks', *Nature*, 183 (1959), 384–5; Stott and Stacey, 'Magnetostriction and Palaeomagnetism of Igneous Rocks', *Journal of Geophysical Research*, 65 (1960), 2419–24.
[65] Cox and Doell, op. cit. (n. 54), p. 655. Stacey (op. cit. [n. 60], t1 s1) recounts that soon after the appearance of the 1960 papers in the *Journal of Geophysical Research* Buddington wrote to him acknowledging the case that Stacey had made and withdrawing the magnetostriction argument.
[66] W. Glen [*The Road to Jaramillo: Critical Years of the Revolution in Earth Science* (Stanford, 1982)] discusses in some detail the rock-dating work at the ANU, the co-operation between Berkeley and the ANU that developed in this area in the late 1950s and early 1960s, and the competition over the reversal time-scale that developed between the Cox-Doell-Dalrymple group at the US Geological Survey at Menlo Park and the McDougall-Tarling group at the ANU. I have borrowed freely from Glen's account except where indicated by references to other sources.
[67] Ibid., pp. 79–81.

[68] Ibid., pp. 81–4 and 210.
[69] J.F. Evernden, interview by H.E. Le Grand, 8 February 1988, t1 s1.
[70] Ibid.
[71] J.F. Evernden and J.R. Richards, 'Potassium-Argon Ages in Eastern Australia', *Journal of the Geological Society of Australia*, 9(1) (1962), 1–49.
[72] Browne's dates were not idiosyncratic; they were often in agreement with or based upon dates assigned by other workers and took into account similarities in petrology and landforms.
[73] This is not to say that the K-Ar dates were unproblematic: several were later 'corrected' in the light of further field studies and field evidence.
[74] Evernden and Richards, op. cit. (n. 71), p. 8.
[75] For a comprehensive discussion of the reversal timescale, see Glen, op. cit. (n. 66).
[76] McDougall, op. cit. (n. 21), t1 s2; for Hess, see Le Grand, op. cit. (n. 1: 1988), pp. 195–201.
[77] McDougall, ibid., t1 s1.
[78] Ibid.
[79] Ibid.
[80] I. McDougall and D.H. Tarling, 'Dating of Polarity Zones in the Hawaiian Islands', *Nature*, 200(1963), 54–56.
[81] Dalrymple (interview by H.E. Le Grand, 9 February 1988) relates that this competition began almost immediately on the publication of McDougall's first time-scale and that it acted as a spur to both groups. The game of leap-frogging each other's published time-scales made it 'a lot more fun'.
[82] A. Cox, interview by W. Glen, 2 May 1978, t4 s1 and t4 s2.
[83] Stacey, op. cit. (n. 60), t1 s1.
[84] Indeed, even if one of these dyads were made to fit, one would probably produce only a description (of dubious accuracy) rather than an analysis of the sort of science practised and the varying receptions that it received. For example, consider 'centre-periphery'. The ANU group and Carey were on the *periphery* of the Australian community of geologists in terms of research programmes, style and theory preferences; however, they were close to the *centre* of the international community if by 'international community' we mean those earth scientists interested in or engaged in controversy over Drift. Within the ANU, there were different *centre* relationships for different aspects of research: with the US (especially the San Francisco Bay area) for K-Ar rock dating and reversal time scale work (McDougall and Tarling), and the UK (especially Cambridge, Newcastle-upon-Tyne and University College London) for palaeomagnetic directionalist studies (Irving). But reversal time scales and polar wandering paths were on the intellectual and institutional *periphery* of geology in the 1950s and early 1960s. Centre-periphery terminology here seems to be mere hand-waving unless one defines the centre *of what* and the periphery *of what*. For example, from the point of view of theory and practice, the ANU group and Carey are at the centre of the periphery, Bullen is at the centre of the centre; and most of the remaining Australian earth scientists were at the periphery of the centre, but this represents a jumble of cognitive, social, institutional and personal factors to which the affixing of centre/periphery labels obscures or misleads rather than clarifies.
[85] See Le Grand, op. cit. (n. 1: 1988), pp. 193–5 and 252–3.

[86] Stacey, op. cit. (n. 60), t1 s2.
[87] Irving, op. cit. (n. 38), t1 s2.
[88] The term 'commonplace' should not be taken to mean that it was on the lips of every geologist or even tacitly accepted by a majority of them. Rather, it is a view that was incorporated in many published accounts.

JAN SAPP

GEOGRAPHIC ISOLATION AND THE ORIGIN OF SPECIES:

*The Migrations of Michael White**

The triumph of genetics is one of the landmarks of twentieth-century science. The institutional and intellectual success of this discipline resulted from many factors. The socio-economic relations between genetics and agriculture, the production of various innovative techniques, including refined microscopes and cytological staining, chromosomal mapping, and the invention of various mutagens such as X-rays and gamma rays, capable of increasing gene mutations at will, were important in keeping genetic research alive prior to the Second World War. The domestication of microorganisms for genetic use and the deployment of techniques borrowed from physics and biochemistry, such as chromatography and X-ray crystallography, were crucial for understanding what genes are and how they control chemical reactions in the cell.[1]

The spread and development of genetic ideas and techniques relied on the migration of individuals between specialities, disciplines and countries. With the rise of experimental biology in Germany during the nineteenth century, biologists from England and the United States frequently visited leading German laboratories to exchange ideas and learn the most recent techniques. By the 1930s, however, the major migration directions had changed. Many biologists began to visit universities and other institutions in the United States where genetics research had flourished since the second decade of the century.

During the 1930s, intellectual migration from Europe to the United States intensified. Many European scientists not only visited the United States to exchange ideas and learn techniques, they immigrated – not by choice, but as political refugees. With the rise of Nazism in Germany and Fascism in Italy, many outstanding scientists of various intellectual persuasions, including Albert Einstein, Enrico Fermi, Felix Bloch, Richard Goldschmidt, Max Delbrück, Salvador Luria and Erwin Schrödinger fled their native countries. Most found homes and places to continue their research in the United States. They subsequently helped to reshape the scientific communities of the New World; indeed, it is

R.W. Home and S.G. Kohlstedt (eds.), International Science and National Scientific Identity, 233–254.

difficult to speak about American science without mentioning their names. How well these scientists assimilated, and the extent to which they helped shape science in the United States, has begun to receive attention by historians of twentieth-century science.[2] That they stayed in the United States after the war may be a measure both of the opportunities they found in their foster country and the difficulties they would have faced in returning to their war-torn countries.

The story of émigré physicists is generally a happy one of assimilation and collaboration in the war effort.[3] However, not all of the German scientists who migrated to the United States between the two World Wars were successful in maintaining the status they previously enjoyed. The experience of émigré physicists contrasts strikingly with that of some leading geneticists such as Richard Goldschmidt and Victor Jollos who migrated from Germany to the United States during the same period.[4]

When understanding the difficulties some scientists had when assimilating into their new culture during the 1930s, '40s and '50s, one has to consider political, cultural, and language differences. With regard to émigré geneticists, at least, one also has to consider difficulties due to intellectual incompatibilities resulting from national differences in intellectual traditions. Unlike European physics, continental genetics was not generally held in high esteem in America. Genetics in Germany developed in virtual confrontation with American genetics. In direct conflict with American genetic orthodoxy, many German geneticists rejected the exclusive role of Mendelian genes and selection theory to account for the origin of species.[5]Moreover unlike the physicists, émigré geneticists did not have a Manhattan Project behind which to rally. Intellectual clashes and social prejudice faced several Jewish refugees from the genetics community of Germany. Richard Goldschmidt, for example, remained outside genetic and evolutionary orthodoxy in the Untied States and was seen by many as an iconoclast. Another German refugee, Victor Jollos, who was given a temporary position in the United States in 1934, was left abandoned by 1941 because of 'racial' prejudice and cultural and intellectual differences.

The war and post-war years in the United States posed great difficulties for many émigré scientists, not simply for those who had migrated by necessity, but also for some who had migrated by choice. After the Second World War, many European scientists longed to leave their war-torn countries and go, as visiting professors or research fellows, to exchange ideas and learn new techniques developed in the United States during the war. Some were given the opportunity to take up residence in the United States. In some cases they, too, then encountered difficulties due to the rise of the Cold War. One who did so was Michael White, and the problems he faced had important consequences for the development of genetics in Australia.[6]

Michael White (1910–1983) was a leading cytogeneticist who began his career working with J.B.S. Haldane in England during the 1930s. He is best known today for his research and books on cytogenetics and problems of speciation. His first book, *The Chromosomes* (1937), went through six editions and was translated into French, Spanish and Italian.[7] In it, he gave a concise and lucid description of chromosome behaviour during mitosis and meiosis, as well as of the relations between chromosome variation and evolution. His cytological work is largely summarized in the successive editions of his now classic book, *Animal Cytology and Evolution* (1945, 1954, 1973).[8] This book was far more ambitious than the first, and represented a survey of all animal cytology in relation to evolution. In his later years, White devoted much of his time to challenging basic assumptions of the 'evolutionary synthesis' of the 1930s and '40s. His criticisms of the orthodoxy that was forming largely concerned mechanisms of speciation. This work culminated in his much-acclaimed book, *Modes of Speciation* (1978).[9]

White's career illustrates various themes in the international relations of science following the Second World War. In 1947, he migrated from England to the United States. Six years later he left for Australia, where he afterwards built a new department of genetics at the University of Melbourne and helped shape the development of genetics research elsewhere the country. In 1964, he became foundation professor of genetics at the University of Melbourne. Genetics was barely a field of activity in Australia when White first arrived there in 1953. This was hardly a place that would be first choice for so distinguished a geneticist. White's immigration to Australia poses two questions of interest in understanding the development and spread of genetic research. How did he come to move to a country where genetics research was so impoverished? How was his research affected by such a move? As we shall see, White's international career, his life on three continents, was the result of various socio-intellectual strategies and counter-strategies.

OUT OF ENGLAND

White received his Bachelor's degree in zoology and human physiology from University College London in 1931. The following year he received a Master's degree and accepted a position as Assistant Lecturer in Zoology at University College. Three years later he was promoted to Lecturer. He remained at University College until 1947 when he was promoted to Reader, but he left for America in the same year. At first glance it might seem difficult to understand why White would leave England where his career had developed so successfully. In order to understand his doing so, one has taken into consideration his pre-university cultural background and the social and intellectual conditions he experienced in England.

White held considerable contempt for polite English society, and in particular its élitist nature. He saw himself as an outsider. Although White was born in London, his early education was in Italy (1914–1920) and France (1920–1927). He acquired a distaste for the class-ridden nature of English society early, and his father had encouraged this attitude. In his autobiography he wrote at length about his father, whom he admired greatly.[10] His father had studied mathematics at Cambridge University in the late nineteenth century, though he never pursued a career as a mathematician. Instead, after living a 'life of a slightly bohemian dilettante and man-about town' in London, he abandoned his bachelor ways, married, and made a modest living by private tutoring before leaving England in 1915. He continued this form of livelihood, making a modest living in Italy and France. He tutored Michael when he had time, and when Michael showed an interest in natural history, his father encouraged it. When his father died of cancer when Michael was 14 years old, he left a written account of his wishes and ideas regarding his son's future. Michael White recalled:

> I was to go to an English University, probably Oxford or, failing that, Cambridge. In the meantime, I was not to be sent to an English 'public' school, even if my uncles and aunts offered to pay for this. Instead I was to enrol with an institution in Cambridge called the University Correspondence College, owned by a certain Mr George Briggs. This organization would train me to pass the necessary entrance examinations. Some money for my university education would became available under my mother's marriage settlement. There were detailed notes as to just what subjects I should study and an assessment of my proficiency in each of them.[11]

White's mother, a poor widow, remained loyal to her husband's dying wishes despite pressure from relatives who wanted to interfere. Michael followed his father's plan almost to the letter. After almost three years at the University Correspondence College, he was prepared to go to University College. He recalled that he did not go to Cambridge or Oxford partly for financial reasons and partly because he did not feel confident about passing the compulsory examinations in Latin and Greek.[12] White felt excluded the moment he entered University College and lived at a men's student's residence in 1927:

> The other students living in this institution were almost all from 'public' schools and with typical British arrogance made fun of the odd boy from continental Europe who had never been to any proper school at all. The only friends I made were some working class lads from the north of England and one student who had been brought up in Berlin because his father was a Diplomat.[13]

White continued:

> Being thin, shy, foreign and very poor, I must have been the least impressive of the students, socially, and the gilded youth of the upper classes suitably despised me. I had almost no social relationships with girl students, many of whom were the daughters of successful medical men with money, clothes and the social position that this implied. In fact, most of my friends during this period were from the faculty of Arts, preparing for careers as teachers in the state schools.[14]

Alas! But something good may have come out of this feeling of isolation. White reflected: 'Lack of money and social rejection by the majority of my fellow students had the effect of making me study much harder in my first year at the University than would otherwise have been the case'.[15] White's autobiography is incomplete; it does not cover the period following the Second World War. However, his wife remembers her husband becoming more and more dissatisfied at that time with intellectual life in England.[16] There was a great deal of infighting among English biologists. And there is no doubt that White himself was engaged in conflict and competition with another of England's leading cytogeneticists, C.D. Darlington.[17]

On the other hand, there were positive reasons for going to the United States where genetics and cytological research were flourishing. White had already acquired a first-hand appreciation of intellectual life in the United States before the outbreak of the Second World War, having spent the academic year 1937–38 as a Rockefeller Research Fellow in the Zoology Department of Columbia University. Columbia University was the home of the original *Drosophila* group that had developed the chromosome theory. It was also the home of the legendary cytologist E.B. Wilson, whose text, *The Cell in Development and Heredity*, was responsible, White claimed, for his becoming a chromosome cytologist in the first place.[18]

From his experience in the United States, White later believed, he had learned to combine two styles of scientific work. As he saw it, the influential figures in England such as Haldane and Darlington were very speculative in their scientific work; they put a premium on theory as opposed to observation. In White's view, they would write 'scientific papers containing brilliant ideas that might or might not be true, after the most cursory, fragmentary laboratory work or none at all'.[19] No one in England, he recalled, taught him 'the need for thoroughness in scientific work'.[20] White began to have a sense of creating his own scientific style at Columbia when he began to observe and work with Franz Schrader and his wife, Sally Hughes-Schrader. Their style, as White saw it, was just the opposite of that of Haldane and Darlington. They were tedious describers who made laborious and detailed investigations and were reluctant to theorize about the meaning of their results. White commented, 'Observation was for them an end in itself and they never accepted Darwin's view that it "must be for or against some view if it is to be of any service"'.[21] White may have deplored the Schraders' descriptive approach to cytology, but he had a great appreciation for the thoroughness of their research. White's strategy for doing good scientific work was to take the thoroughness of the Schraders' style and combine it with the brilliant theorizing of the Haldane or the Darlington.

During the war years, White worked at the British Ministry of Food and was unable to do any research. He returned to University College in October 1944 and, in spite of 'material difficulties and a rather heavy teaching load', managed to get a few pieces of research completed.[22] When he could not do research during the war, he spent his time writing his book, *Animal Cytology and Evolution*. This book integrated a veritable encyclopedia of information on animal chromosome cytology into evolutionary biology. It had an enormous influence on cytogenetic and evolutionary thinking and soon became the Bible of those interested in the cytology of animal chromosomes and the evolutionary pathways leading to the rich diversity of chromosomal phenomena. White pioneered a new field of evolutionary and cytological research. Following the lead of the Russian émigré, Theodosius Dobzhansky, he initiated a program of 'population cytology'. Instead of examining the chromosomes of single individuals, White studied whole series in a population. The individual variations he detected in such population samples often provided him with important clues concerning the evolutionary course of chromosomal changes. In sum, *Animal Cytology* formed an important part of the convergence of disciplines during the 1930s and '40s that shaped the evolutionary synthesis. William Atchley gives the following evaluation of the book:

This work, the first critical account of cytology since E.B. Wilson's 1925 monograph on *The Cell in Development and Heredity*, provided an encyclopedia of information on animal chromosomal cytology. However, it would be highly misleading to label *Animal Cytology* as simply an enormous compilation of cytological observations, although it obviously was that. Rather, the most important aspects of this book was that it established evolutionary cytology as a discipline in its own right. *Animal Cytology*, along with Dobzhansky's *Genetics and the Origin of Species*, and Simpson's *Tempo and Mode in Evolution*, brought together aspects of cytology, genetics evolution, systematics and paleontology and integrated them into what could truly be called a synthetic theory of evolution.[23]

White quickly became well known following the appearance of *Animal Cytology*. The year following the publication of his book he arranged another visit to the United States. Like many other European scientists, he wanted to see something of the work that had been done there during the preceding eight years. He again applied to the Rockefeller Foundation for financial assistance.

The Rockefeller Foundation played a leading role in providing aid to foreign scientists who wished to visit the US for short periods to renew contacts and in many cases to obtain stimulation to carry on work that had been interrupted during the war period. The Foundation had a continuing policy of following the careers of its past Fellows and in helping them whenever possible, particularly in the case of those who had already distinguished themselves in research. The publication of White's influential new book provided him with enough credibility to warrant support as one of the first European scientists to be awarded a Rockefeller Foundation Fellowship after the war. The Rockefeller Foundation Fellowships were given with an important stipulation: that Fellows who visited the United States return to their native countries when the term of the fellowship was completed.

On this visit to the United States, White had no intention of settling down and working in a single laboratory as he had done in 1937–38. That did not appeal to him for, as he explained to one of the Rockefeller

Foundation officials, he could do that equally as well at University College.[24] White developed a strategy to match the synthetic and broadly based nature of his book of 1945. His plan for this visit was to spend three or four months travelling to various important genetics centres in the United States.

White arrived in the United States from England via Canada in June 1946 and stayed until late September. His tour was extensive. He visited laboratories at Harvard and discussed cytological and genetics problems with America's outstanding cytogeneticist, Barbara McClintock, and others at Cold Spring Harbor. He visited the laboratory of Edward Tatum at Princeton to learn of work on the biochemical genetics of *Neurospora* and bacteria that had been developed during the war years. He also visited the laboratories of leading geneticists at Austin, Texas. He spent a week in New York City with members of the Zoology Department of Columbia University.[25] He also spent long hours in discussion with the superb Russian émigré geneticist, Theodosius Dobzhansky. Dobzhansky's book, *Genetics and the Origin of Species*, published in 1937, played a prominent role in the construction of the evolutionary synthesis of the 1930s and 1940s. As Carson writes, nowhere was its effect greater than in cytogenetics: 'Dobzhansky wrote from a background of knowledge of genetic and chromosomal variability, both within and between species'.[26] As mentioned earlier, White's book, *Animal Cytology*, which helped to provide an evolutionary framework for integrating cytological work, had been deeply informed and inspired by Dobzhansky's work.

White profited greatly from his trip to the United States, both intellectually and socially. Indeed, the relationships he made in his visit to the United States in 1946 proved very important for his future career, for in the following year, White accepted a job as professor of zoology at the University of Texas. The genetics group there was strong and active. It included such well known geneticists as Wilson Stone, Robert Wagner and J.T. Patterson. Dobzhansky's book had led to a new speciality of genetics work dealing with cytogenetics, evolution and natural populations. Much of this work had been pioneered at the University of Texas by Patterson, Stone and Griffith.[27] The addition of White to the group working on grasshoppers made the University of Texas the major centre for cytogenetic investigations of natural populations.

White began doing outstanding and innovative work on the cytology of populations of midges and the evolutionary cytology of *Orthoptera*, which had some advantages over *Drosophila* for field studies. He had some good graduate students including several from Canada.[28] He and his wife Sally settled into Austin and assimilated into American culture well.[29] In fact, by 1951, the Whites were already in the process of taking out citizenship in the United States. However at that very moment disaster struck. Four years after arriving in Austin, the Whites found themselves in a terrible situation and confronted with the possibility of deportation.

RED, WHITE AND BLUE

In 1951, during the period of McCarthyist hysteria, the Texas legislature passed a so-called 'anti-subversive law'. As applied to White and other government employees, the law stated that no state funds could be paid to (a) any person who had ever been a member of any communist organization or (b) any person who had at any time in the previous ten years been a member of any of the organizations named as subversive by the Attorney General, or (c) any person who had worked as a registered lobbyist for the Soviet Union or any of its satellites. Anyone refusing to sign an oath that he or she was not then and never had been a communist could not be paid from funds covered by the bill.

White could not take such an oath. From 1932 until the beginning of 1935, he had been a member of a student organization at University College affiliated with the Communist Party in Great Britain. This was a time, with the great depression and the rise of Nazism in Germany, when many European scientists and other intellectuals had sought refuge in the Communist Party. Sally White, who had also been a member of the Communist Party, in Cambridge, remembers that, unlike herself, her husband later felt ashamed of having ever been a member. White later recalled the atmosphere of those early years:

> This is something that I have much regretted in later years as a kind of blot on my intellectual record. But as an unemployed or virtually unemployed, young scientist in 1932, faced with the rise of Hitlerism in Germany and with the evidence of this in the refugees beginning to arrive in England, it seemed a natural enough thing to do. I believe that I was never a very convinced Communist. Virtually from the start I was thoroughly mistrusted by the more dedicated and fanatical members of the groups who regarded me as a rather unreliable member and potentially a 'Deviationist'.[30]

White remembered[31] that as a young student in his early 20s, he had faithfully attended a series of seminars on Marxist philosophy led by the famous crystallographer, J.D. Bernal. Bernal had exercised a great influence over young intellectuals at Cambridge and London.[32] White read the main works of Lenin but recalled feeling very ambivalent about them. While leading biologists such as J.B.S. Haldane became more and more active in the Communist Party, White found himself to be less than devoted. After three years he broke with the student communist group and in 1935 subsequently joined the British Labour Party of which he was a member until the outbreak of war in 1939.

Although White later regretted his relationship with the Communist Party, it was not 'a blot on his intellectual record' in the 1930s. It certainly became one, however, with the rise of McCarthyism in the United States. To White and many of his colleagues in 1951, it looked like professional ruin for him. Immediately, White's academic friends in the United States attempted to come to his rescue. Many hoped that during the course of the year the law would be declared unconstitutional. In the meantime, it was clear to several of them that White needed to leave Austin for at least a year. Immediately, they wrote to Warren Weaver of the Rockefeller Foundation for help.

Approaching the Foundation for help was a good idea. During the 1930s and 1940s, the Rockefeller Foundation had greatly fostered the development of genetics, especially the integration of biochemical and physical approaches with traditional genetics ones. Founded in 1913 and reorganized in 1928, the Rockefeller Foundation developed a system of research grants for individuals and projects, and mastered the art of conducting a large programme of relatively modest grants. As Kohler[33] has pointed out, the organization and style of the Foundation's programme acted as a model for federal science agencies, such as the National Science Foundation, formed in the United States after the Second World War. From 1932, Warren Weaver was the director of the natural sciences division of the Rockefeller Foundation. His principal aim was to develop molecular biology. In fact, it was Weaver himself who coined the expression 'molecular biology' in 1938.[34] Weaver was a former physicist and by the early 1930s, genetics seemed ripe to him for exploiting the techniques of chemistry and physics.

As mentioned above, the Rockefeller Foundation officials took great interest in providing funds for international co-operation among scientists. The scope of the Foundation's interest extended well beyond the borders of the United States. It played an active role in helping to organize and support the work of leading European biologists, especially in France, Britain and Germany. Rockefeller Foundation officials followed the careers of their past Fellows and sought advice from them about what work in their speciality was important and which young researchers were most promising and worthy of support.

The American geneticist, George Beadle, first brought White's problem to Weaver's attention. Beadle was one of the leaders in the new domain of the biochemical genetics of microorganisms that developed during the Second World War.[35] Prior to Beadle's lead, biochemistry and genetics had for the most part gone separate ways in their development. Beadle's work with the biochemist Edward Tatum on the bread-mould *Neurospora* was one of the first and most successful attempts to merge genetics with biochemistry. The Rockefeller Foundation had followed Beadle's career closely. In fact, Rockefeller Foundation officials accepted special responsibility for bringing Beadle and Edward Tatum together in the late 1930s. After the Second World War, with Rockefeller Foundation support, Beadle developed a major programme of biochemical genetics at the California Institute of Technology that developed in close relationship with the biochemistry division led by Linus Pauling, who worked on the structure of proteins. In 1958, Beadle shared a Nobel prize with Edward Tatum and Joshua Lederberg.

By 1951, White had become recognized as one of the few outstanding cytogeneticists in the world. Beadle had learned of White's political situation from one of White's colleagues at the University of Texas, Wilson Stone. It seemed to Beadle and Stone that the best strategy for helping White was to provide a fellowship for him to leave Texas. White had intended to spend part of the following summer at the California Institute of Technology. Beadle and some of his colleagues there were willing to take White on for a year if the Rockefeller Foundation would help. However, Beadle believed they would run into trouble with a 'certain extreme group of trustees' who would have to be told the full story. The question was not so much one of money, but of lending the Rockefeller Foundation's authority to persuade some of the trustees of the California Institute of Technology to see the justice in their actions. Beadle explained the situation to Weaver:

Mr. M.J.D. White, Professor of Cytology at the University of Texas, is in an impossible situation which I suggested to Wilson Stone you might be able to help on. It seems he was a Communist Party member back in the 30's. He became convinced that this was all wrong in 1936 and quit the party. After this he apparently actively opposed Communism. The Texas legislature has tacked onto the appropriations bill a rider providing that anyone refusing to sign an oath that he or she is not now and never has been a Communist cannot be paid from funds covered the bill. You can see where this puts White. It looks like sure ruin professionally either way for him. It looks to Stone and I agree, that the only solution is for White to leave either temporarily until the law, which I should think would certainly not stand up, could be changed, or permanently.

We come into the picture this way: White had intended to spent part of next summer here and we are enthusiastic about this plan. Now Stone wonders if we could take him for a year (leave of absence basis). This Sturtevant, Watson and I (who know the whole story) and the rest of the Biology Division (which does not know about the former c. party connection because we want to keep it quiet if possible) think this is a fine idea. We are short on cytology and we believe White would fit in excellently here. The catch is not so much funds, although now it is so late that all budgets for next year are made up, but rather the probable reaction of the Trustees who Sturtevant, Watson and I agree would have to be told the whole story. Lee is already on a rather hot spot on the Communism question with a certain extreme group of trustees – fortunately a small one. Ernest feels that if an outside agency were to propose that we help in what is obviously a very unfair situation to White there is a good chance that our Trustees would be willing to O.K. an appointment here ...

The question is simple – do you see any way to help?[36]

Weaver replied promptly, wasting no words in informing Beadle that he was not willing to stick his neck out in support of White:

I do not know Mr. White. I have no information whatsoever concerning his general reliability. I have no basis whatsoever for convincing myself of what I assume is probably the case – namely, that he has completely abandoned his communistic ideas. But it would certainly be only under very extraordinary circumstances, where I had personal convictions based on direct personal experience and evidence, that I would feel justified in exerting the slightest amount of pressure or suggestion. To do so in cases of this sort is a questionable and dangerous procedure at best; and to do so when I have no real basis for doing so seems to me, I am sorry to say, quite impossible.[37]

Beadle wrote back stating that on reflection he sympathized with Weaver's attitude. Beadle himself was going to talk over the White issue 'unofficially' with one of the more liberal trustees of the California Institute of Technology. If he advised against recommending White, Beadle and his colleagues were willing to let the matter drop. Nonetheless, Beadle emphasized to Weaver that he was convinced of White's political convictions and lamented the risks of losing an outstanding cytogeneticist:

I have confidence that the facts are as stated although, of course, it's never possible to be completely certain. I hope a way can be found to keep White in this country. Otherwise we lose a top-notch man in a field where there are far too few.[38]

At Beadle's suggestion, Wilson Stone, who knew the situation best at Texas, also wrote to Weaver. Stone wrote an impassioned plea, not in order to obtain some institutional authority, but only in an attempt to secure a Rockefeller Foundation Fellowship for one year for White. 'The price of liberty is eternal vigilance', he wrote,

> but this implies an understanding of the concept of liberty so that we can know how to defend it. A retroactive law that damns all men for actions which carried no onus at the time or were perfectly innocent of mal-intent does not, in my opinion, defend liberty.[39]

Stone explained to Weaver how White had broken with the Communist Party, had worked for some years with the British Labour Party, 'and in fact, found them too far to the left, so that for a long while he might be regarded as a liberal middle-of-the-road man'. Stone pointed out that a fellowship would allow White to rewrite his book, *Animal Cytology and Evolution*, and continue his research on the cytology of grasshopper populations. Stone hoped that in the course of the year the law might be changed. If, however, it was not, and the oath was upheld as legal, the fellowship would give White an opportunity to find a suitable job elsewhere. Finally, Stone emphasized to Weaver that White could provide evidence that 'he broke with and opposed the communists even in the war years when such an attitude was unpopular'.

Subsequently, during the late 1940s, many leading communist biologists, provoked by the rise of Lysenkoism in the Soviet Union, broke with the Communist Party. Backed by Stalin in 1948, Lysenko and his supporters had denounced Mendelian genetics, cytogenetics and gene theory as Fascist, bourgeois and idealist theory that was incompatable with dialectical materialism and Soviet ideology. All genetic research carrying a Mendelian bias was actively rooted out of biology in the Soviet Union.[40] Anti-genetics propaganda in the Soviet Union was soon matched by active anti-Lysenko propaganda in America. In 1950, the Genetics Society of America organized a major symposium to counteract Lysenko propaganda and defend the theoretical and practical benefits of genetics. Supported by funds from the Rockefeller Foundation, the programme celebrated the advances in genetics that had occured over the previous fifty years.[41] The resulting text, *Genetics in the 20th Century*, brought together essays of some of the most distinguished geneticists of the day. White was one of the leading participants. His contribution was an historical survey of 'Evolution and Cytogenetic Mechanisms in Animals'.[42] Stone knew all too well of the anti-communist hysteria that was being used to eliminate left-wing and other critical persons from a wide range of institutions. With regard to White, he recommended that if there was trouble at the California Institute of Technology, he might go to Cold Spring Harbor or Columbia University instead. Weaver again responded unfavourably. He was not interested in the historical context of White's Communist ties, he was more concerned with the contemporary hysterical context. Weaver wrote to Stone that he simply was not able to provide a fellowship for White:

> There are two primary reasons why we could not consider favourably an application from him. The first reason is that the funds which were available this year for fellowship appointments are exhausted. In the second place we have an invariable rule here that the fellowship awards are given only to individuals who have full-time positions, from which they can obtain a leave of absence for the period of the fellowship, and to which they are expected to return at the completion of the fellowship. Therefore, Dr. White would not be eligible, if he has to leave the University of Texas.[43]

White did not go to the California Institute of Technology. Instead, Sally White recalls[44] that they spent the next year at Cold Spring Harbor where her husband was a guest researcher. They returned to Austin the next year, but by 1953 they were forced to leave Texas and migrate to Australia.

GEOGRAPHIC ISOLATION AND SPECIATION

White's move to Australia was initiated by Dobzhansky at Columbia University. Dobzhansky was a good friend of both White and the geneticist Otto Frankel, Chief of the Division of Plant Industry of Australia's Commonwealth Scientific and Industrial Research Organisation (CSIRO). It was with Frankel's help that White arrived in Australia. Frankel had been born in Australia and trained in Berlin in the early 1920s under the celebrated geneticist Erwin Bauer. He recalls[45] that he came to dislike Germany which was becoming increasingly right-wing and where he experienced a great deal of anti-Semitism. In 1929 he migrated to New Zealand to take up an appointment as geneticist at the Wheat Research Institute in Christchurch. He remained there for twenty-three years doing economically and theoretically important wheat breeding work. He was elected to the Royal Society of London in 1953 for work done in New Zealand. Frankel moved to Canberra in 1951 when he accepted the position as Chief of the Division of Plant Industry.

Frankel had known Dobzhansky for some time. Dobzhansky had been his guest in New Zealand. Frankel remembers that Dobzhansky wrote to him in 1953, asking if he could find a place for White because he had to leave the United States. He was going to be deported otherwise.[46]

At the time Dobzhansky wrote, Frankel was engaged in building up genetics research at the CSIRO. How he got White into Australia was a closely guarded secret at the time for, as Frankel recalls, 'there was McCarthyism in Australia as well'. Frankel had met White in 1939 at the John Innes Institute, Merton, England. However, they had not associated. By the early 1950s, White was well-known for his cytological research and texts. Under normal circumstances, Frankel recalls,[47] he would not have gone out of his way to get White. There were more obvious people whom he would have wanted to recruit. After all, White's research was on animal cytology, not on plants.

In order to understand the reasons for Frankel wanting to accommodate White, one has to take into consideration the state of genetics in Australia, and at the CSIRO in particular, where there was little genetics and no cytology before Frankel arrived. It was natural that Frankel would want to establish a strong group of geneticists. Indeed, it was expected of him. Genetics elsewhere in Australia was also extremely weak. There were only two other, newly-established geneticists in the country. At about the same time as Frankel arrived in Australia, David Catcheside was appointed professor of genetics at the Waite Agricultural Research Institute affiliated with the University of Adelaide. The Chairman of the CSIRO, Ian Clunies Ross, had also appointed J.M. Rendel, a *Drosophila* geneticist from Edinburgh, to establish a CSIRO unit for animal genetics at the University of Sydney. Frankel was regarded as the third one.

The Division of Plant Industry was a rather 'applied' institution before Frankel arrived. The CSIRO had begun with agricultural research in the late 1920s. The Division of Plant Industry, which Frankel now headed, was one of the earliest Divisions formed and became the largest. It had been largely concerned with applied research and Frankel's brief was, as he put it, 'to introduce science to it'. It was this brief that justified Frankel having an insect cytologist as a research fellow. To have a zoologist in a plant research institute was rather unconventional to say the least; but Frankel recalls that Clunies Ross 'took this in his stride'.

That White had been a communist as a student did not bother Frankel, for he himself had also been a one in his student days. An idealistic young man, strongly communist orientated, he had wanted to do something practical about world hunger and hence decided to study agriculture. It was this ambition that had led him to work under Erwin Bauer in Berlin. Frankel spoke to Clunies Ross and explained White's predicament to him. Clunies Ross was sympathetic. Frankel recollects that 'Clunies Ross was a generous person, for him not to support someone who was being persecuted by McCarthyism would be unthinkable'. They agreed to offer White a three-year fellowship. However, in doing so Frankel suspected they were taking considerable risks. In fact, the only person who knew what was happening, besides Frankel and Clunies Ross, was Frederick White, CSIRO's deputy chairman.[48] Indeed, the three had taken a considerable risk in accommodating White for the CSIRO had been plagued by McCarthyism.[49]

Frankel recalls that White's years at the CSIRO were good ones. They liked him very much, he was 'good scientific company' and a most valuable colleague. During his three years at the Division of Plant Industry, White spent considerable effort learning the natural history of, and investigating spermatogenesis in Australian grasshoppers. In 1954, a second edition of his *Animal Cytology and Evolution* was published. When his fellowship expired in 1956, White accepted an appointment as professor of zoology at the University of Missouri. Frankel recalls the discussions over his leaving Australia. White said he 'wanted to go home' to the United States. He claimed he could not get an education for his children in Australia.

As it happened, a year was not over before White wrote to Frankel seeking to return to Australia. And Frankel succeeded in finding a position for him. While White had been in the United States, the professor of zoology at the University of Melbourne had died. By the time White wrote to Frankel, applications for the vacant chair had closed but no appointment had been made. Clunies Ross was on the University Council and Frankel asked him if the appointment could be reopened. Clunies Ross obliged and Frankel wrote to White encouraging him to apply.

White returned to Australia in July 1958 and remained there for the rest of his career. Immediately he began to develop an active genetics research programme. By March 1959 he was seeking Rockefeller Foundation support for his main project on chromosome inversions in grasshoppers under field and laboratory conditions. He had five graduate students but, like others in Australia, lamented the difficulty of attracting young graduates into biology. Between this time and 1964, he devoted a great deal of energy to establishing a separate department of genetics at the

University of Melbourne. When this new department was created in 1964, White became its first professor and he occupied the chair until his retirement in 1975.

White thus became geographically isolated from his American colleagues. There is no evidence, however, that this hindered his intellectual development, though it may have been important in shaping its direction. Between 1962 and 1968, White began to deviate from the orthodox views regarding evolutionary processes, and his distance from centres of genetics and evolutionary research in the United States may have helped him to develop his own style. The most important challenge he issued to established evolutionary dogma concerned modes of speciation and the role of geographic isolation.

By the mid-1960s, evolutionists, including White himself,[50] had generally reached a consensus that new species of sexually reproducing animals arose exclusively through geographic isolation. During the late 1960s and early 1970s, White began to challenge this view and argue that there was no one mechanism of speciation but several, depending on the 'life style' of the organism. Geographic isolation was not always a necessary prerequisite for speciation. White eventually brought his ideas together in his well known book, *Modes of Speciation* (1978).

Certainly, White's interest in the problem of speciation did not originate in Australia. As Ernst Mayr has noted,[51] White's earlier work on 'population cytology' got him interested in the role of chromosomes in speciation. Yet there was a uniquely Australian component that contributed greatly to his still controversial theories of speciation, for these were based on cytogenetic work on chromosomal rearrangements in an exclusively Australian group of grasshoppers. White had first turned his attention to the effects of chromosomal inversions on viability and size in a small grasshopper, *Moraba* (now commonly known as *Vandiemenella*) in 1953. After studying certain species of this genus for many years, he found anomalies concerning the evolution of chromosomal rearrangement that did not fit well with orthodox assumptions about speciation. White found that closely related species differed sufficiently in chromosomal structure for this to act as a cytogenetic isolating mechanism leading to the emergence of new species. On this basis, he developed his new theory of 'stasipatric speciation' according to which chromosomal changes may initiate the evolution of new species. White's work on *Moraba* was crucial to his new theory of stasipatric speciation.[52]

Though White's geographical isolation from his colleagues overseas may have been important for the intellectual development of his theorising on speciation, he himself did his best to minimize it. One of White's first graduate students, John Martin, recalls[53] that from the moment White arrived at the University of Melbourne, he did his best to maintain close links with the United States and to internationalize genetics research in the university. In fact, according to Martin, there were more American post-doctoral fellows in the Genetics Department during White's period at the University of Melbourne than ever before or since. Although there was an active group of biologists working in the field of radiation genetics at Melbourne before White arrived, through his influence genetics built up great momentum and also changed direction.

White's influence on the development of genetics in Australia extended well beyond the borders of the University of Melbourne. Australia soon became a major centre for cytogenetic studies of speciation. According to Martin, during the 1960s an active group of biologists at the Research School of Biological Sciences at the Australian National University began to follow White's lead on problems of speciation. White played a major role in establishing the group at Canberra and in the development of a Department of Population Biology there in the 1960s. In fact, he played an instrumental part in the appointment of its first professor, Bernard John, an English cytogeneticist. Perhaps somewhat ironically, the research of the group at Canberra developed in direct conflict with the research and theorizing led by White in Melbourne. Healthy, heated debates ensued throughout the 1960s. When White retired, he spent the last years of his research career as Visiting Fellow at the Department of Population Biology, Australian National University.

CONCLUDING REMARKS

This brief glance at the career of Michael White brings into relief various issues concerning the migration of scientists and the development of genetics in Australia. White's arrival in Australia was not planned. It has to be understood as resulting from strategies and counter-strategies in a larger historical context of international migration of scientists following the Second World War. White came to Australia by an indirect route; he left England where he felt excluded and migrated to the United States, lured by the thriving centres of genetics research there. Although White was intellectually compatible with his peers in America, cultural differences soon caught up with him. In McCarthyist America he was persecuted because of his past political associations; Australia became his place of exile.

In passing we have noted that Otto Frankel likewise moved to Australasia for negative reasons. Like White, he was an outsider in the country in which he was trained as a scientist. Yet while Frankel had later been invited to move from New Zealand to Australia to develop genetics research, White's arrival was the result of a political secret. To Frankel it was almost comical that White, a zoologist, worked in the Division of Plant Industry at the CSIRO.

Yet his remaining in Australia after the hysterical period of McCarthyism was over was not automatic. Australia, under-developed and weak in genetics research and geographically distant from genetics centres elsewhere, was not an attractive place from White's point of view. There was no organized attempt to create a suitable position for him at one of the universities. It was again only by accident that the chair of zoology fell vacant at Melbourne, giving White the opportunity to foster the development of genetics in Australia. Once inside the university, White could act purposefully. He was successful in building a Department of Genetics and in attracting postdoctoral fellows and visiting scholars from overseas. He thus minimized any intellectual distance that might have resulted from his geographical distance from genetics centres overseas.

White's local institutional successes were matched by his research in the international academic marketplace. White had brought his procedures, skills and problems in population cytology with him to Australia. Yet his new Australian environment played an important part in modifying his views on speciation. Indeed, the origin of White's theory of stasipatric speciation cannot be understood without reference to the Australian group of grasshoppers that helped to shaped it.

Department of History and Philosophy of Science
University of Melbourne
Australia

NOTES

* I am grateful to the archivists of the Rockefeller Archive Center and to Doug McCann and the Australian Science Archives Project for help in collecting documents. I also thank Otto Frankel, Sally White and John Martin for their personal recollections, and Rod Home for his editorial expertise. This work was supported with

funds from the Australian Research Grants Scheme.

[1] On the development of genetics, see for example, G.E. Allen, *Life Sciences in the Twentieth Century* (Cambridge, 1975); G.E. Allen, *Thomas Hunt Morgan. The Man and His Science* (Princeton, 1978); C.E. Rosenberg, *No Other Gods* (Baltimore, 1976); Robert Olby, *The Path to the Double Helix* (London, 1974): H.F. Judson, *The Eighth Day of Creation* (New York, 1979); Jan Sapp, *Beyond The Gene: Cytoplasmic Inheritance and the Struggle for Authority in Genetics* (New York, 1987).

[2] See for example, Donald Fleming, 'Emigré Physicists and the Biological Revolution', in Donald Fleming and Bernard Bailyn, eds., *The Intellectual Migration: Europe and America, 1930–1960.* (Cambridge, Mass., 1969), pp. 152–189; Charles Weiner, 'A New Site for the Seminar: The Refugees and American Physics in the Thirties', ibid., pp. 190–233. Nathan Reingold, 'Refugee Mathematicians in the United States of America, 1933–1941', *Annals of Science*, 38 (1981), 313–338.

[3] Weiner, op. cit. (n. 2) p. 226 summarizes the experience of refugee physicists concisely when he writes:

> The refugee physicists, who came from the centers of physics research in Europe, were able to move freely in their new environment, adding their knowledge and individual research styles to work under way at American institutions, or seeking out new fields for exploration. Their opportunity to contribute was enlarged because a large number of them had been carefully placed by American colleagues at institutions where they were needed and could be most productive.

[4] On Richard Goldschmidt, see Scott Gilbert, 'Cellular Politics: Goldschmidt, Just, Waddington and the Attempt to Reconcile Embryology and Genetics' in R. Rainger, K.R. Benson and J. Maienschein, eds., *The American Development of Biology*, (Philadelphia, 1988), pp. 311–346. On Jollos, see Sapp, op. cit. (n. 1), pp. 60–65.

[5] On the difference between American and German approaches to heredity between the two world wars, see Sapp, op. cit. (n. 1), pp. 54–86, Margaret Samosi Saha, 'Carl Correns and an Alternative Approach to Genetics: The Study of Heredity in Germany Between 1880 and 1930', Ph.D. dissertation, Michigan State University, 1984; Jonathan Harwood, 'The Reception of Morgan's Chromosome Theory in Germany: Interwar Debate Over Cytoplasmic Inheritance', *Medizinhistorisches Journal*, 19 (1984), 3–32; Jonathan Harwood, 'Genetics and the Evolutionary Synthesis in Interwar Germany', *Annals of Science*, 42 (1985), 279–301.

[6] There is, as yet, little secondary literature devoted to the study of the career of Michael White. For brief accounts of his contributions to cytology and evolutionary theory, see William R. Atchley, 'M.J.D. White: The Scientists and the Man', in William R. Atchley and David Woodruff, eds., *Evolution and Speciation: Essays in Honour of M.J.D. White* (Cambridge, 1981), pp. 3–20, and Ernst Mayr, 'Michael James Denham White', *American Philosophical Society Yearbook*, 1984, pp. 156–159. A doctoral dissertation on White's life and work is being planned by Doug McCann at the University of Melbourne.

[7] White, *The Chromosomes* (London, 1937).

[8] White, *Animal Cytology and Evolution* (Cambridge, 1945).

[9] White, *Modes of Speciation* (San Francisco, 1978).

[10] White, 'A Life in Three Hemispheres', unpublished autobiography, 1978–1983, pp. 18–45. White Papers, Series 1, File 1/3, University of Melbourne Archives.

11 Ibid., p. 45.
12 Ibid., p. 48.
13 Ibid., p. 57.
14 Ibid., p. 62.
15 Ibid.
16 Telephone conversation with Mrs. Sally White, Melbourne, 14 April 1988.
17 See Atchley, op. cit. (n. 6), p. 6.
18 White, op. cit (n. 10), p. 106.
19 Ibid.
20 Ibid.
21 Ibid., p. 107.
22 M.J.D. White to H.M. Miller, 31 January 1946; Series 401D, Record Group 1.1, Rockefeller Archive Center, Tarrytown, New York.
23 Atchley, op. cit. (n. 6), p. 7. For similar views concerning the place of White's *Animal Cytology* in the 'evolutionary synthesis', see Hampton Carson, 'Cytogenetics and the Evolutionary Synthesis', in Ernst Mayr and William B. Province eds., *The Evolutionary Synthesis* (Cambridge, Mass., 1980), pp. 86–95; at p. 91.
24 White to H.M. Miller, 26 March 1946; Series 401D, Record Group 1.1 Rockefeller Archive Center.
25 White to G.P. Pomerat, 4 October 1946; Series 401D, Record Group 1.1 Rockefeller Archive Center.
26 Carson, op. cit. (n. 23), p. 92.
27 Ibid.
28 Wilson Stone to Warren Weaver, 5 February 1951; Series 400D Record Group 1.1 Rockefeller Archive Center.
29 Mrs. Sally White, telephone conversation 14 April 1988.
30 White, op. cit. (n. 10), p. 90.
31 Ibid., pp. 90–92.
32 See Gary Wersky, *The Visible College* (New York, 1978).
33 Robert Kohler, 'A Policy For the Advancement of Science: The Rockefeller Foundation, 1924–29', *Minerva*, 16 (1978), 480–515; p. 481.
34 Robert Kohler, 'The Management of Science: The Experience of Warren Weaver and the Rockefeller Foundation Program in Molecular Biology', *Minerva*, 14 (1976), 279–306.
35 For a detailed discussion of the origins of biochemical genetics, see Jan Sapp, *Where the Truth Lies: Franz Moewus and the Origins of Molecular Biology* (New York, 1990).
36 Beadle to Weaver, 15 May, 1951; Series 401D, Record Group 1.1, Rockefeller Archive Center.
37 Weaver to Beadle, 18 May 1951; Series 401D, Record Group 1.1, Rockefeller Archive Center.
38 Beadle to Weaver, 24 May 1951; Series 401D, Record Group 1.1, Rockefeller Archive Center.
39 Stone to Weaver, 19 May 1951; Series 401D, Record Group 1.1, Rockefeller Archive Center.
40 On the rise of Lysenkoism in the Soviet Union, see Z.A. Medvedev, *The Rise and Fall of T.D. Lysenko*, translated by I. Michael Lerner (New York, 1971); David

Joravsky, *The Lysenko Affair* (Cambridge, Mass., 1970); Robert Young, 'Getting Started on Lysenkoism', *Radical Science Journal* 6/7 (1978), 81–106; and Richard Lewontin and Richard Levins, 'The Problem of Lysenkoism', in Hilary Rose and Steven Rose, eds., *The Radicalization of Science* (London, 1976), pp. 32–65.

41 On the reactions to Lysenkoism in the United States, see Sapp, op. cit. (n. 1), pp. 163–191.

42 White, 'Evolution and Cytogenetic Mechanisms in Animals', in L.C. Dunn, ed., *Genetics in the 20th Century* (New York, 1951), pp. 333–369.

43 Weaver to Stone, 25 May 1951; Series 401D, Record Group 1.1, Rockefeller Archive Center.

44 Sally White, telephone conversation, 14 April 1988.

45 Frankel, telephone conversation, 14 April 1988.

46 Dobzhansky let the Rockefeller officials know about White's fate and how much he was missed at Texas. H.M. Miller recorded in his diary after meeting Dobzhansky at Columbia University: 'In commenting on the situation at Texas, D. Lamented the great loss of past NSA fellow M.J.D. White, the brilliant British cytologist who left recently for a position at Canberra, Australia. He said that Otto Fraenkel, Austrian-born, brilliant geneticist who had spent 20 years in New Zealand is now head of plant research at Canberra' (H.M. Miller, 'Diary', entry for 16 October 1953, Series 401, Record Group 2, Rockefeller Archive Center).

47 Frankel, telephone conversation, 14 April 1988.

48 Ibid.

49 See, Jean Buckley-Moran, 'Australian Scientists and the Cold War', in Brian Martin, C.M. Ann Baker, Clyde Manwell and Cedric Pugh, eds., *Intellectual Suppression: Australian Case Histories, Analysis and Responses* (North Ryde, NSW, 1986), pp. 11–23.

50 See Atchley, op. cit. (n. 6), p. 12. White's first, seminal paper showing a change in his view appeared in 1968: White, 'Models of Speciation', *Science,* 159 (1968), 1065–70.

51 Mayr, op. cit. (n. 6), p. 157.

52 White saw his views on speciation, based upon studies of flightless insects, to be in direct conflict with those of Ernst Mayr who, as a result of his experiences with birds, has emphasised the geographical component in speciation processes. See White, 'Scientific Contributions of M.J.D. White', unpublished 4-page manuscript (1979–80?), p. 1 (White papers, University of Melbourne Archives). Differences in organisms studied by biologists have often had profound effects in shaping the production of conflicting biological theories; see Sapp, op. cit. (n. 1), especially pp. 232–234.

53 Personal communication.

STEPHEN J. PYNE

ANTIPODAL FIRE:

Bushfire Research in Australia and America

TWO FIRES

In the beginning was the fire;
Out of the death of fire, rock and waters;
And out of water and rock, the single spark,
the Divine truth.

– Judith Wright, 'The Two Fires' (1955)

In January 1939 – Black Friday – bushfires savaged southeastern Australia. Perhaps 1.4 million hectares in Victoria were burned, some 71 lives were lost, and damages went beyond the capacity of the economic calculus of the day to assess. With the majority of the fires set by humans and nearly all of them beyond the capability of humans to contain, the Black Friday holocaust publicized a massive breakdown in rural fire practices. Bushfires were of course nothing alien to Victoria; its establishment as a crown colony in 1851 had been accompanied by a panorama of fires that exceeded those of 1939.[1] But the bush was then lightly occupied by squatters and diggers. When Black Friday sent a smoke pall that blackened even the skies over Tasmania, the land had been occupied by Europeans for a century. A Royal Commission, headed by Judge Leonard E.B. Stretton, was called upon to survey the wreckage and recommend remedial measures.[2]

But the fires of Black Friday, awesome as they were, paled before a flame kindled later that year in Europe as Nazi Germany invaded Poland – the onset of a global conflagration that would sweep over Australian society. As the Imperial Japanese Army captured Hong Kong and bore down on Singapore in December 1941, Prime Minister Curtin issued his famous declaration that 'without any inhibitions of any kind, I make it quite clear that Australia looks to America, free of any pangs as to our traditional links or kinship with the United Kingdom'.[3] In February 1942 Broome and Darwin were bombed and burned in apparent preparation for

R.W. Home and S.G. Kohlstedt (eds.), International Science and National Scientific Identity, 255–287.

an invasion of the Australian continent. Australia exchanged its traditional military alliance with Britain for one with the United States. The American alliance would lead, in the post-war years, to a friendly but sometimes smothering 'invasion' of American ideas and popular culture.[4]

Together the response to these two events shaped the modern era of bushfire protection and research in Australia. Stretton minced no words in indicting Australia's peculiar climate and its feckless rural fire practices as the source of the disaster. The old ways could not continue. Similarly the realization that Britain could no longer sustain Australia – that Australia would have to seek military alignment with America instead, that it would have to bolster its miniscule (by Asian standards) population, that it was time to reassess Australian identity in terms of other than as a member of the British Empire – announced a new era of Australian history. A massive reconstitution of Australian society was under way that would both shape and be shaped by a new program of bushfire protection.

What makes the story particularly compelling is that the two trends were not always complementary. The imperative to seek out a suitable Australian bushfire strategy did not always equilibrate with the emerging American alliance and its heady transfer of American ideas and institutions Down Under. The reason is that the American system of fire protection, which redoubled after the war, and the Australian system, which was emerging in the war's aftermath, were founded on fundamentally different axioms.

The American strategy based fire protection on a programme of fire control. It stressed rapid detection and initial attack, and after the Second World War it quickly mechanized and intensified. Scientific research moved into laboratories, shaped into mathematical models that were informed by analytical principles – a thrust into Big Science that took fire research far beyond its origins in forestry. By contrast, the Australian strategy based fire protection on fire use. It proposed an alternative strategy of hazard reduction burning – the routine, controlled, light burning of fuels so that wildfires, which were considered inevitable, would burn with less intensity. Scientific inquiry stressed field studies over laboratory experiments, the empirical over the analytical, and management applications over intellectual curiosity.

The concepts behind the Australian strategy did not rise spontaneously out of the ashes of 1939. 'Burning off' was a common practice in rural

Australia. But, animated by the spectre of Black Friday and sanctioned by the stature and vivid rhetoric of Judge Stretton, it soon acquired organization, conviction and legitimacy. In the two decades that followed, the Australian fire community would discipline those concepts into an institutional reality and sustain them with a programme of scientific research. Controlled burning was, however, a concept that Americans had previously subjected to ferocious debate and repudiated. Consequently, even as the Australian-American alliance matured and broadened, the fire protection strategies of the two nations drifted towards friendly but apparently irreconciliable antipodes.

RED CENTRE: AUSTRALIA AS A FIRE CONTINENT

A debate about appropriate strategies of bushfire protection would be trivial if the bush were not important to Australian history and if fire were not an important feature of the bush. But Australians have always tied their identity to that of the bush, and the bush has burned with powerful consequences and relentless regularity. Fire is the latest, the most subtle and the most malleable of the environmental processes that have shaped the Australian bush out of the ancestral rainforest that was the biotic legacy of Gondwana.[5] Through its interplay with humans, in particular, fire established itself as the red centre of Australia.

The bush evolved under the influence of three primary determinants. The first was a relentless erosion and leaching that left most of the continent's soils deficient in phosphorus and trace nutrients. Then, during the Tertiary – as Australia drifted northward into the tropics and became isolated from other Gondwana fragments – came the Great Upheaval. Australia experienced drought. Aridity appeared as both a seasonal and a secular phenomenon. The degradation of soils had been incessant but quiescent; drought, however, could arrive suddenly, it heralded other disturbances, and it occurred with irregular rhythms. The Australian biota had to adapt to episodic dryness and atonal patterns of disturbance, among them fire. Rainforest was increasingly crowded out by Scleroforest – scleromorphic vegetation that included trees, shrubs and grasses. Organisms that had been marginal to Rainforest became central to Scleroforest. The Great Upheaval inspired a Great Inversion of Australian vegetation.

By the end of the Pleistocene, however, another revolution was in progress. Swiftly the last of the Rainforest was driven into relict sanctuaries, while within the Scleroforest one genus, *Eucalyptus*, rapidly assumed dominance. The rate of change was astonishing, remarkably recent, and catalyzed by fire. It is no coincidence that the revolution coincides with the arrival on the island continent of the genus *Homo*, outfitted with bold firesticks.

Fig. 1. Major Thomas Mitchell meets the Bogan tribe. Mitchell's guide and the Bogan headman both hold firesticks. Reproduced from T.L. Mitchell, *Three Expeditions into the Interior of Eastern Australia*, 2 vols. (London, 1839).

Unlike other organisms, humans could control ignition. They could alter the timing and frequency of fire, vary its intensity, and realign its geography; they could reconstitute the fire regimes of Australia. There is compelling evidence that *Homo* has done precisely this. The environmental conditions that favoured the eucalypt also favoured the Aborigine; their ranges were coextensive and within those ranges, their densities were correlative. The land the Europeans settled, as Sylvia Hallam has put it, 'was not as God made it. It was as the Aborigines made it'.[6] The relationship between fire and Aborigine was reciprocal. 'If fire was maintained by the Aborigines', as Phyllis Nicholson writes, 'it is also true that the Aborigines were maintained by fire ... The evidence that fire was the indispensable agent by which Aboriginal man extracted many of his resources from the environment is irrefutable'.[7] Rhys Jones has referred to the semi-cultivation of the landscape by Aboriginal fire as 'fire-stick farming'[8] (Figure 2).

Fig. 2. Aborigines using fire to flush out kangaroos. Reproduced from Joseph Lycett, *The Natives and Scenery of Tasmania and N.S. Wales* (1817).

The European, too, was a fire colonizer, and he introduced another dimension to the geography of Australian fire. Not only could the European vary ignition patterns, he could attack, directly and indirectly, the fuel complexes on which fire fed. From Europe also came traditions of burning that had been disciplined into agricultural cycles and circumscribed by traditions of fixed land ownership. These practices were transferred, often incompletely and eccentrically, to Australia. Broadcast fire was associated with nearly all forms of agriculture. The 'firestick habit in Australia', lectured E.H.F. Swain in 1927, 'commenced with the aborigine'. But the torch was soon transferred to the European pioneer – graziers, selectors, miners. It 'became a white man's habit', and reached such dimensions that

at this stage the rural population is so enamoured of the use of fire that wherever he now goes, and in the periods of highest fire hazard, instead of in the winter time, the bushman blithely distributes his matches, and the schoolboy is learning to follow in father's footsteps.[9]

'Burning off' became itself a principal means of fire protection. Rural Australia ensured common protection against bushfire by protective firing on a colossal scale, both in prepared fuelbreaks and among the surrounding bush. Otherwise, neighbors helped neighbors as needed – an expression of mateship captured in Henry Lawson's famous poem, 'The Fire at Ross's Farm' (Figure 3).

Fig. 3. Fighting the bushfire – boughs, buckets and bags. The bushfire was a great s piece of Australian settlement. Reproduced from *The Town and Country Journal* (23 December 1871).

Under the European onslaught, bushfires doubled, even quadrupled in number, swelled in intensity, and swept over the landscape – one of the worst and most impressive of the environmental plagues that accompanied the European expansion across Australia.[10] A dramatic increase in charcoal represents an increase in fuels, which reflects the immensity of land-clearing. Especially large fires broke out in Victoria in 1851 (Black Thursday) and Gippsland in 1898 (Red Tuesday).[11] The Black Thursday fires burned perhaps 76,000 km^2. When compounded with other biotic manipulations and exceptional aridity, such as the Long Drought of the 1890s, the results could be devastating.

The Great Upheaval of the Tertiary was being replayed, with greater intensity, in the Holocene. Native vegetation was uprooted, rearranged, replaced with breathtaking rapidity. Exotic flora and fauna reconstituted whole regimes, in some places increasing and in other diminishing the prevalence of fire. New sources (and purposes) from fire came into play. Escape fires, careless burning, and outright incendiarism were abundant, part of the unsettling of Australia, the rapid and often violent reconstruction of the Australian landscape. It is no accident that the bushman – the transient, the digger, the bushranger, the swagman, the sundowner, the rural migrant – became the great Australian type.[12] Fire was an inevitable by-product of the friction between a restless European and a tindered Australia. Only with time did a new equilibrium of fire use and fire control come into play. Then the introduced fire plague became, like rabbits and white clover, an endemic, disciplined exotic.

TOUCHED BY FIRE: AMERICA AS A FIRE CONTINENT

Until the twentieth century the American experience with fire had not been qualitatively different.[13] Lightning kindled fires routinely, particularly in areas that experienced at least seasonally dry periods. The American Indian burned habitually. The tallgrass prairies were only the most spectacular instance of a biome sustained by anthropogenic fire; when the fires were removed, the landscape quickly reverted to forest. Europeans, too, adopted native fire practices and then added others; exotic flora and fauna were introduced, often to run wild; fire was ubiquitous on the frontier and in agricultural settlements, massive wildfires were common, and protective burning was a routine technique for fire control. In 1904 surveyors for the Plumas forest reserve in northern California described how

> To the casual observer, and even to shrewd men ... the fires seem to do little damage. The Indians were accustomed to burning the forest over long before the white man came, the object being to improve the hunting ... The white man has come to think that fire is a part of the forest, and a beneficial part at that. All classes share in this view, and all set fires, sheepmen, and cattlemen on the open range, miners, lumbermen, ranchmen, sportsmen, and campers. Only when other property is likely to be endangered does the resident of or the visitor to the mountains become careful about fires, and seldom even then.[14]

Substitute 'Aborigine' for 'Indian' and 'bush' for 'forest' and the passage could describe Victoria as fully as California.

Yet there were important differences between America and Australia. The American scene was more varied, pliable, and resilient. While important, and in some biomes vital, free-burning fire tended to be one of a competing ensemble of processes that shaped the American landscape. America had experienced geological and biological rejuvenation. Wildland fire, although dreaded in its more violent forms, was accommodated; the belief existed that it would recede before the inexorable course of settlement, like Indians and bison and other manifestations of the wilderness. Some even imagined an America largely purged of fire. Temperate America could be farmed – its fuels transformed – in ways that subtropical Australia could not; its population was much greater, more widely dispersed over a variety of fire regimes; and industrialization spread sooner and more intensively.

In Australia, fire went beyond the status of a vital process to become an informing principle. Fire had touched America; it branded Australia. There seemed to be no escape. Virtually nowhere in Australia was spared routine fire.[15] Tropical Australia was burned annually.[16] Fire visited the arid interior more episodically, though emphatically, dictated by the pattern of winter rains, not summer drought. In 1974–75 as much as 15% of Australia burned in one season, most of it in the interior.[17] The Mediterranean environments of Australia experienced long dry summers and frequent fire. Climatologists have noted that 'potentially bad fire seasons' tend to occur about once every three years, 'bad fires' every six or seven years, and 'very bad fires' every 13 years.[18] The circumstances in southeastern Australia were particularly intense. Here cold fronts drafted desiccating air out of the interior deserts that dried out fuels and drove fire to the southeast; then came a wind shift – the southerly buster – that drove fire to the east along multiplying fire fronts.[19] Tasmania was more heavily loaded with fuels but otherwise it, too, was trapped in the same meteorological matrix that made southeastern Australia into a colossal fire flume. It is estimated that nearly half of the economic damages inflicted by bushfires occur on 3% of the continental surface.[20] The richest soils, the greatest botanical profusion, the heaviest human settlements – all crowded into the fertile crescent of the southeast. It was a landscape that humans would have to share with fire.

That accommodation was difficult. Compared to American fire, Australian fire was more ruthless, relentless, concentrated, singular. Broadcast fire was the great tool of settlement and the hearth fire was a warm reminder of 'Home', but bushfires conveyed an alien presence that brooded over the landscape. The smokey pall, the crackling flames of a thousand tiny fires, the threat of an explosive rush out of the bush – the spectre was always there, oppressive and threatening. In America, following European precedents, fire was associated with wolves, and it was conceivable that both could be extirpated. In Australia bushfire was more often identified with snakes. There was no possibility of eradication. Bushfire would thrive as long as the bush existed.

In Henry Lawson's famous story, 'The Drover's Wife', that association sets the emotional tone of the tale. A black snake escapes into cracks in a bush barky. The woman – her husband gone a-droving, the farm incapable of supporting a family in droughty times on poor soils – must guard her children alone. She passes the night in a chair waiting for the snake to emerge. She relives all the oppressiveness of life in the bush; her first memory is of a bushfire that threatened the selection until neighbors arrived to help. At last the snake worms out, the woman and her dog attack and kill it, and she dumps the dead body onto the hearth fire. But the vigil is endless. There will be more snakes.

Like them, wildfire always waited in the bush. It dominated the dynamics of the Australian landscape as the eucalypt dominated the composition of Australian forests and woodlands. Yet the susceptibility of the bush to fire and the ability of humans to control fire made it an especially seductive, almost addictive tool. Like the eucalypt and the Aborigine, the European became a fire creature, entangled in a faustian bargain with fire, able to tolerate bushfire only by adding more fire to the bush. There seemed to be no escape from an endless spiral of burning. Bushfires emanated from the core of the continent. With increasing fervor, the red centre controlled Australia's green perimeter.

SOUTHERN ANTIPODE: THE AUSTRALIAN STRATEGY

> Men who had lived their lives in the bush went their ways in the shadow of dread expectancy. But though they felt the imminence of danger they could not tell that it was to be far greater than they could imagine. They had not lived long enough. The experience of the past could not guide them to an understanding of what might, and did, happen.
>
> – Judge Stretton, *Report of the Royal Commission to Inquire into the Causes of and Measures Taken to Prevent the Bush Fires of January, 1939*

European fire could, in theory, be 'naturalized' along with European flora and fauna. In those areas of Australia where the bush had been converted into a Neo-Europe, something like this had occurred. Routine burning-off and ad hoc assistance among neighbors could quell normal bushfires. Against the occasional conflagration, however, there was no response. The outcome was a peculiar fatalism. Because normal fire years could be accommodated through laissez-faire folk practices, there was little incentive to install a formal bushfire protection system. but because there was nothing that could counter a full-blown holocaust, there was similarly little movement to create rigorous institutions for fire control. Either way, bushfires were here to stay. 'No worries, mate'.

Not all of rural Australia, however, was given over to graziers, nomadic cocky farmers, and migrants. The Empire also brought forestry from Europe. Plantation forestry became a significant presence in the Mediterranean climates of Australia. Elsewhere reserves of native forests were established with increasing frequency and turned over to professional foresters for management. In both instances, bushfire was a problem that could not be blinked or burned away. Throughout the Europeanized globe, foresters were the great antagonists of free-burning fire.[21] Forestry meant more than technology transfer: it communicated a moral energy, a political fervor that transcended nationalist boundaries and the size of the lands under its direct supervision. Not their number but their claim to the high ground, the intensity of their commitment, and their belief that all elements of forestry, including fire, were amenable to scientific analysis gave foresters a major role in determining fire policy and programmes.

Gradually strategies were developed within the Empire to cope with fire and the social practices that sustained it. Canada argued for an aggressive programme of fire suppression that paralleled American practices. But India and Africa, more syncretic, were equally influential as potential exemplars for Australia. The Australian colonies, later States, were eclectic in their concepts, and it fell to Western Australia to develop the most coherent programme. The catalyst was Sir D.E. Hutchins, an Empire forester trained by Sir Dietrich Brandis himself and tested by fire in India, Cape Colony, and east Africa. In 1914 he toured Australia and vigorously promoted the classic Empire strategy – the preparation of detailed 'working plans' that called for clear-felling, a site burn, fire protection without grazing, and then fire protection with grazing.[22] The formula was adopted in Western Australian forests, then under the direction of C.E. Lane-Poole, another graduate of the Empire Forestry School at Oxford. After Lane-Poole left to direct forestry for the Commonwealth, the youthful Stephen Kessell enthusiastically advocated these methods.[23] Western Australia, late to be settled, came to display the most rationalized of forestry and fire protection systems. But while Western Australian foresters laid out an institutional matrix, its conceptual core was less firm.

Between the Royal Commission that reported on the Victorian fires of 1939 and the Royal Commission that surveyed the Western Australian fires of 1961, the essential framework of an Australian system of bushfire protection was hammered out, legitimated and put into the field. Its core was a rationalization of burning off, not surprising when it is realized that Australian forests, native and plantation both, tended to be islands in a burning sea of grass and eucalypt litter. Forests were absorbed within that larger dominion of extensive rural landscapes, and their overseers accommodated the fundamental rural fire protection strategy, burning off. It was a programme for extensive rather than intensive land use, suitable for a lightly populated continent dedicated to coarse commodity production. The 1939 fires exposed the fatal liabilities of ad hoc programmes – their reliance on folk practices and rural practitioners, the indiscriminate use of fire as a universal solvent, the operational limitations. By the early 1950s, under the impress of foresters, those practices were being reconstructed into a uniquely Australian strategy of bushfire protection.

The first expression of the new regime appeared in Western Australia when foresters there consciously decided that hazard reduction burning on a massive scale was the only cost-effective means by which to establish fire protection. Fire use had never been abandoned or condemned outright, but what the new policy did was to insist that the forest burning actually be conducted as prescribed and that it be widened to encompass virtually all the State forests in need of protection. The 1961 Dwellingup fires, and the Royal Commission headed by G.R. Rodgers to investigate the disaster, confirmed that strategy.[24] Those areas that had been subjected to hazard reduction burning on a regular schedule had survived the fires with minimal damages; those areas that had not been recently fired under controlled conditions were destroyed by wildfire.

What escalated the programme beyond Western Australia, however, was the work of two foresters from New South Wales: Alan G. McArthur and R.H. Luke. A native of western New South Wales (the son of a wheat farmer), a graduate of the national forestry school in Canberra, an intensely practical man who viewed 'North American' exemplars with suspicion, and a successful, often cynical bureaucrat – McArthur rationalized rural fire practices into an interdependent system, legitimated the use of broad-acre burning for hazard reduction by means of scientific and historical studies, and successfully reconciled this reformation with the undercurrent of a renewed Australian nationalism.[25] From his positions with the Commonwealth Forestry and Timber Bureau, and later the Division of Forest Research of the Commonwealth Scientific and Industrial Research Organization (CSIRO), he sponsored considerable bushfire research and helped co-ordinate fire studies conducted by other CSIRO divisions. He even installed the subject into the curriculum of the Australian National Forestry School in Canberra. McArthur rationalized folklore and folk practices into something like a science. The 'red steer', as bushfires were commonly described, became an object of empirical study. By the mid-1960s the Australian system – what can only be termed an institutional and conceptual revolution – was accomplished.

The Australian strategy makes a number of suppositions. It assumes that large fires are inevitable under conditions of Australian fuels and climate. It assumes that rural (and ex-urban) environments will continue to rely on volunteer bushfire brigades. It assumes that Australian biotas have been subjected to periodic fire for eons and that the critical species and ecosystems have evolved adequate adaptations to surface burning. It accepts as a truism that a reduction in fuels will result in a diminution of fire intensity and fire damages. It asserts, in an almost syllogistic conclusion, that hazard reduction is the only practical strategy upon which to base bushfire protection.

Between McArthur and A.R. King, a physical chemist with CSIRO, a research agenda was designed to support this programme.[26] That strategy sought to identify the environmental controls on fire behaviour and to codify that knowledge in ways that would be meaningful to practitioners. It searched out ways to protect property and firefighters from fires, most spectacularly in the development of a portable fire shelter. It massed evidence to vindicate prescribed fire as an environmentally sound technique – by demonstrating the manifold fire adaptations of the genus *Eucalyptus*, the domain of which is practically coextensive with the regions in need of protection; by determining the rates of actual fuel build-up and the combustibility of Australian fuel beds; by establishing the fundamentally innocuous properties of bushfire smoke on human health; by developing technologies through which immense quantities of land could be safely burned on the scale and with the frequencies demanded; and by documenting the tens of thousands of years during which Australia had been fired by Aborigines.

Between them, McArthur and King mobilized the primary scientific institutions in Australia with an interest in fire. When McArthur assumed his directorships, he gave bushfire research real bureaucratic clout. Similarly they established a research style. They conducted case studies of large fires and applied laboratory analysis to special concerns such as the combustion of eucalypt leaves and litter. But mostly McArthur used field trials – literally hundreds of experimental fire – to measure how fires spread in real-world fuel beds within the range of environmental conditions most relevant to Australian circumstances. It was a research strategy that clearly grew out of rural experience in burning wheat stubble, cane fields, paddocks and forested bush.

What McArthur did for folk knowledge, R.H. Luke did for folk practices. McArthur took great pains to design and package his research into forms usable by practitioners. But as much as anyone Luke rationalized rural fire protection – not only burning off, with which he sympathized, but actual firefighting as well. More than anyone else, Luke brought order to an emerging system of volunteer bushfire brigades. He wrote manuals, oversaw enabling legislation, urged that local control remain at the core of rural fire protection.[27] After they retired, the two grand old men of the Australian strategy collaborated on a standard text, *Bushfires in Australia* (1978), that synthesized their collective experience.

It was the genius of the Australian strategy, as promulgated by McArthur and Luke, that it emerged out of practices of long antiquity and that it accommodated the realities of bushfire and rural life in Australia. If the McArthur research agenda had to fight off North American models, the Luke scheme for bushfire brigades had to resist British exemplars, particularly the Britain that nationalized its fire services in 1947. What resulted was a unique synthesis: volunteer brigades, financed in good part by private and public insurers, integrated through country (or rural) fire authorities, but executed by local ruralites, supported by a practical field-based programme of scientific research. Limited resources – money, equipment, personnel – meant that an all-out fire suppression programme on the model of metropolitan fire services or after the example of North American forestry services was impossible. But the emerging Australian strategy argued equally that such an approach was unnecessary and unworkable. Fire control would supplement, not supersede, fire use.

NORTHERN ANTIPODE: THE AMERICAN STRATEGY

All this stood in striking contrast to developments in America. The Australian and American strategies had diverged in ways that were too fundamental to gloss over or reconcile through simple technology transfer. The American strategy had been forged by a series of events that occurred early in the twentieth century: the transfer of vast public forest reserves to the U.S. Forest Service (in 1905), the traumatic fires of 1910, the light-burning controversy that flared into public debate at exactly the same time, the Weeks Act that established the terms for a Federal-State alliance in co-operative fire protection (the basis for a national system), the spirit of young professionals inflamed by the reformist spirit of the Progressive Era and the conservation crusade, the call to arms of President Theodore Roosevelt for a life of 'strenuous endeavor' and of philosopher William James for a 'moral equivalent of war' to be waged against the uncontrolled forces of nature. All these – along with greater demographic pressures and the drive to specialized land use and the general decay of rural America – shifted control over fire practices from a broad rural base, grounded in folk traditions, to a more specialized political bureaucracy staffed by professional foresters. By the onset of the Second World War, this transfer was more or less complete.[28]

Its intellectual dimension was expressed in the light-burning controversy that pitted a farrago of 'folk' practitioners of fire use against a more selective band of foresters who argued that fire protection could only be based on fire control. Light burning advocated the routine use of low-intensity ('light') fires as a means of reducing fuels and thereby eliminating the possibility of high-intensity conflagrations. Fire protectionists countered that, at best, controlled burning was a stop-gap measure, that any fire was damaging, that fire exclusion was the only ultimate outcome that could be tolerated. Most of the protectionists were foresters with the U.S. Forest Service. In effect, light burning, which was less a system than an alliance of Forest Service critics, sought to orchestrate the common fire practices of the frontier and accommodate them to an industrializing society.[29]

But frontier fire practices were countered by European exemplars and, partly because of the timing of the 1910 fires, the more nativist strategy lost out. The fight was bitter, dragging out through the 1920s. The intensity of the light-burning advocates, however, forced American foresters to sharpen their arguments for an aggressive strategy of fire control, a strategy first shaped in California where it was promoted as 'systematic fire protection'. As written by Coert duBois, chief of the national forests in California, systematic fire protection expressed the values and methodologies of the industrial engineering of its day, Taylorism, applied to the problems of fire control.[30] The essence of the approach was the unquestionable fact that every fire could be controlled if it were detected and attacked quickly enough. What systematic fire protection did was to elaborate – with mind-numbering rigour and statistical detail – just how a forest service could install lookout towers, patrols, crews, trails and roads, telephone lines and so on to accomplish that end.

As the debate entered into newspapers and general magazines, the light-burning controversy became less a dialogue than a dialectic. It wound down, painfully and full of lingering anger, after an official panel of foresters appointed to review the issues formally condemned light burning as a basis for fire protection.[31] The debate acquired considerable political significance because the Weeks Act (1911), bolstered by the Clarke-McNary Act (1924), had made fire protection the basis for Federal-State co-operative programmes in forestry.[32] Anything that questioned Forest Service policy compromised the de facto nationalization of forestry.

As foresters, the protectionists believed they spoke with the backing of scientific authority. The architect of the co-operative fire strategy, Chief Forester William Greeley, had proclaimed as early as 1911 that 'firefighting is a matter of scientific management, just as much as silviculture or range improvement'.[33] Fire control operations and studies on the effects of fire were part of the initial agenda set by the Service when, in 1916, it established a Branch of Research. Most research had strong connections to operational needs, and much of what passed for research consisted of field trials and statistical analysis conducted by rangers with line responsibilities, not by a cadre of professional scientists. The methods were those of silviculture and forestry economics. Not surprisingly, the field trials were intended more as demonstrations than as true experiments. If they failed to confirm the expected results, their data were suppressed.[34]

By the 1930s, the conservation programmes of the Roosevelt Administration bestowed enormous resources on forestry, and camp after camp of Civilian Conservation Corps enrollees was directed into fire suppression and pre-suppression projects. The enlargement of means at hand stimulated an enlargement in thinking about the ends to which these means might be put. Where systematic fire protection had been confined to sites of obvious economic value, its domain limited by available resources, it was now extended across the continent. The Forest Service adopted its '10 a.m. Policy', which called for the suppression of every fire – in every locale, in every fuel type – by 10 a.m. in the morning following its initial report or, failing that, by 10 a.m. the succeeding day. While advertized initially as an 'experiment on a continental scale, the promulgation quickly shed its tentative origins and became dogma. It was a strategy that could only be imagined in a context of virtually unlimited resources.

Instead of returning after the Second World War to the *status quo ante*, however, the swelling of American affluence and the liberation of war surplus materiel allowed the system to intensify and project itself into such distant realms as the interior of Alaska. By the mid-1950s, firefighting was mechanized – most spectacularly by the conversion of Second World War bombers into aerial tankers by which to deliver water and chemical retardants to active fires. Americans propagandized on behalf of the U.N.'s Food and Agriculture Organization (FAO) and wrote texts on fire control for a global audience. Two national fire prevention programmes – Keep America Green and Smokey Bear (Co-operative Forest Fire Prevention Campaign) – continued after the wartime emergency passed. Fire consciousness was at a fever pitch. America, it seemed, was on its way to a role as global fireman to match its self-anointed status as global policeman. Light burning remained anathema. To Americans light burning was unnecessary and conceptually suspect, a professional embarrassment. Advocates of light burning belonged with perpetual motion mechanics and circle-squarers.

The postwar era thus re-emphasized the core of the American system. Wartime experiences had only reinforced its informing concept, rapid detection and suppression – 'force enough, fast enough'. When war surplus materiel became available in large quantities, particularly after the Korean conflict, firefighting mechanized almost overnight. Its most spectacular expression was the conversion of TBMs and other military aircraft to fire-retardant bombers, the adaptation of helicopters to fire control operations, and the creation of a rapid deployment force in the form of specialized, highly mobile 'shock troops', the inter-regional fire suppression crew. The resources available for fire protection seemed to increase annually. Forest Service hegemony went virtually unchallenged in areas of policy, logistics, expertise and research.

A similar reformation affected fire research. Long before Australians began devising fire-danger meters, Americans had constructed similar devices for every region, their values empirically derived from field trials in select fuels and, particularly, by the statistical processing of tens of thousands of fire reports. The world war and its cold war successor intervened, however. Galvanized by a new liaison with the military – the Second World War had been a fire war, which ended with the explosion of a spectacular new incendiary, the atomic bomb – fire research joined in the acceleration of American science, moved beyond the intellectual confines of forestry, and relocated from the field to the laboratory.

By the mid-1950s it seemed possible to mimic the style of Big Science, to establish specialized laboratories, to merge field trials on large fires with 'analytical' solutions encoded into fire behaviour models. The agenda of fire research would be set more by the physics of fire behaviour than by the field needs of forestry. Military and civil defence interest in atomic warfare promised to fund the transition. The National Academy of Sciences – National Research Council established a Committee on Fire Research to oversee the new investments in fire research and a series of heady 'correlation' conferences were sponsored to identify conceptual concerns and to map out institutional boundaries. It seemed possible to create an all-purpose fire model that, being founded on first principles of physics, would transcend particular fuels and regions. By 1961 the Forest Service had established three fire laboratories; within a decade, a mathematical model for predicting fire behavior had emerged from the laboratories and was put at the core of computer models for nearly all aspects of fire management. Not only in strategy but in choice of scientific methodology, the American and Australian systems had diverged.

THE AMERICAN INVASION AND THE AUSTRALIAN RESISTANCE

Australian foresters had long studied American examples, and Australians knew far better what was happening in Canada and America than Americans knew of events in Australia. In the postwar period the liaisons between Australia and America, strengthened by the wartime alliance and by a de facto postwar protectorate, expanded into many fields. American forestry, and through it American strategies for wildland fire protection, joined the general cultural 'invasion' of postwar Australia by American institutions, businesses, ideas and popular culture.

In 1951 two Australians, one of them R.H. Luke, joined an FAO 'fire study tour' to America. Subsequently, similar tours were arranged on a routine basis between North Americans and Australians; every six years a group from one continent would journey to the other, and these visits would be staggered such that tours were conducted, one way or the other, every three years. In the late 1960s, Australia joined America, Great Britain, and Canada in large-scale field experiments on mass fire known as Project Euroka, the Australian counterpart to Project Flambeau. The trials were intended to test the behaviour of very large fires, the kind anticipated to occur in the aftermath of a nuclear war. With massive landclearing underway in conjunction with the Snowy Mountains Scheme, Australia could offer sites for field trials, and some burns were conducted at Tumut.[35] All this occurred as the Australian system was coming into definition.

Yet the architects of the Australian strategy successfully resisted the American Way. In part, their resistance reflected a deeply held scepticism that a North American-style bushfire protection scheme could succeed under Australian conditions. In part, too, it expressed a stubborn if inchoate Australian nationalism. With ties to Britain fraying and linkages to America uncertain, Australians had to redefine themselves in new terms and with new relationships. Predictably, they turned to the bush – to the indigenous flora and fauna that Australia alone possessed. But any redefinition of Australian identity in terms of the bush would inevitably mean a redefinition of fire practices. The creation of the Australian strategy became part and parcel of that larger cultural process.

The consequences radiated beyond the island continent. What began as a unilateral transfer soon evolved into a more lively exchange. Beginning in 1962, the privately endowed Tall Timbers Research Station in Florida inaugurated a series of widely promoted annual conferences in fire ecology at which Australians became active participants. It is important to understand that these conferences occurred outside the American forestry establishment. They instigated a global survey of fire practices, with a particular interest in the promotion of controlled burning. Through them Australian bushfire research acquired international visibility, and the evolving Australian strategy posed an important counter-example to developments in North America.[36]

Fig. 4. The American strategy flying high – a B-17, converted to service as an air tanker, drops retardant on the Wenatchee fires, Washington State, 1970. Photo courtesy U.S. Forest Service.

As a capstone, and in self-conscious contrast to North American predilections for 'water bombers', Australians invented 'fire bombers' in the mid-1960s as a method of igniting tens of thousands of hectares in the name of hazard reduction. Fire bombers, water bombers – the dichotomy between the Australian and American systems could be symbolized by these two contrasting technologies (Figures 4 and 5).[37]

Fig. 5. The Australian counter-example – a Cessna 337 drops incendiary capsules in support of an aerial prescribed burn near Bemboka, 1968. Courtesy A. Edwards.

The Australian system was thus an undertaking of considerable intellectual achievement and courage. The Australian strategy directly challenged the better publicized American strategy. That example was even carried forcibly back to America, as American foresters came under intense pressure to use fire more widely – to introduce fire into natural landscapes like wilderness, and to use fire to promote biologically sane approaches to wildland management, burning fuels as a means of reducing the intractable quantum of large fires that continued to burn despite ever-increasing investments in firefighting hardware and crews. When, in grudging acceptance of these criticisms, American foresters converted to a philosophy of prescribed fire in the early 1970s, Australian colleagues felt not only vindication but exhilaration.

There is little doubt that the publication of global counter-examples such as Australia had influenced the debate about American fire practices. Whatever other groups had capitulated before the Americans, the Australian forestry and bushfire community had held its ground. The old lament that Australians were followers, lucky outcasts, happy parasites and expatriated aliens, ever calloused towards and biased against their native land, had been challenged by the architects of the Australian system. That was the good news.

WILD BUSH, URBAN BUSH

> Where Sydney and the Bush meet now,
> There is no common ground.
> – Les Murray, 'Sydney and the Bush (1976)

The bad news was that even as the Australian system came into definition, the assumptions on which it rested were being challenged. The same pressures that had worked in America to reform fire practices also worked in Australia. But where those pressures forced Americans to accept some version of controlled burning, they compelled Australians to improve their fire suppression capability and modify their single-minded usage of fire. At issue was not a strategy of fire use alone, but the character of Australian society and the patterns of land use within which a fire programme had to operate.

An explanation begins with postwar demographics.[30] Between the end of the Second World War and the Ash Wednesday fires of 1983, Australia's population had more than doubled, from 7 million to approximately 16 million. A good portion came from a postwar 'baby boom', not unlike that which occurred in Canada and America. But much came from a deliberate effort to promote immigration. Britons, however, did not revive the Australian connection to the extent that Australian planners had expected; most of the European immigrants came from eastern and southern Europe and, consciously defining itself as an 'Asian' country, Australia began admitting Asians, collectively promoting an ethnic diversity and accepting problems of cultural assimilation that the country had never before experienced. Most of the 'New Australians' clustered in urban centres. Equally, there was a steady erosion of rural industries, except for mining, with an internal migration from Outback to metropolis. And within the city-states there was a partial out-migration of affluent Australians to the 'urban bush', far-flung suburbs nestled in 'natural' settings like the Adelaide Hills, the Blue Mountains, and the Dandenongs.

The assumptions of the Australian strategy about extensive land use and the suitability of broad-acre treatments, about rural fire lore and practices, about a rural population familiar with fires use and sympathetic with burning by professional élites – all were subverted by the population boom of New Australians who, as a whole, were ignorant about Australian bushfires. It is no coincidence that the Australian strategy was promulgated by Australians who had grown up during the longest period in their national history without sustained immigration. Now, even those who moved into the urban bush came from metropolitan environments; most had never experienced personally a major bushfire or participated in traditional 'burning off' rituals. Moreover, they brought with them new attitudes about the bush itself and their desired relationship to it.

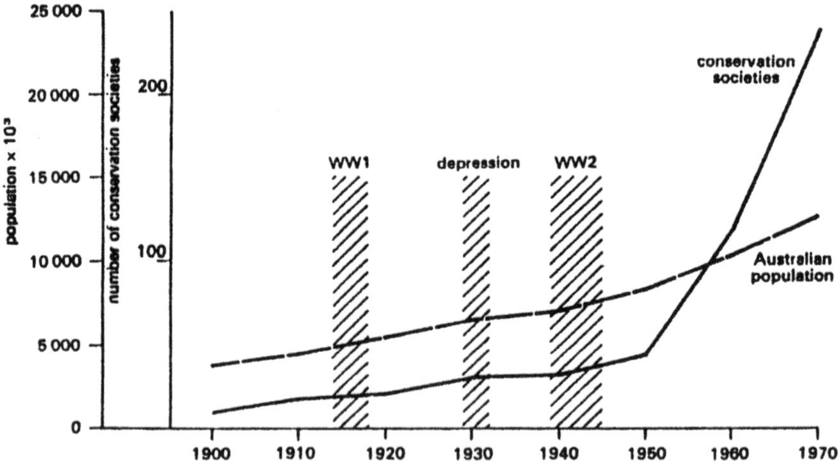

Fig. 6. Increase in number of conversation societies in Australia in relation to the general population. Reproduced from *Report of the Committee of Inquiry into the National Estate* (1974).

The era coincided with a new style of environmentalism that swept across Europe and America. Some concepts and activist concerns were transferred intact to Australia from overseas. Others developed from indigenous sources, as Australians re-examined their relationship to the bush. The desire to define and preserve 'native' Australian flora and fauna, interest by post-industrial societies in 'wild' nature in general, hostility towards alien organisms, suspicion of public forestry, distrust of industrial land uses – all animated an environmental movement in Australia.[39] The growth of conservation societies far outpaced the growth of the general Australian population (Figure 6). The New Australians seemed to want something – what, even they perhaps weren't sure – that was less Euro-centric; they wanted to accept Australia, and themselves, as something other than Neo-Europes; they wanted something beyond homogeneity. To complement Australia's new ethnic pluralism they argued for a new environmental pluralism.

The Australian strategy thus became an object of public controversy. The situation was truly antipodal. In America, environmental groups thundered for more fire in the landscape – preferably 'natural', but almost any fire would do. In Australia, environmentalists wanted restrictions imposed on the practice of broad-acre burning. Aerial ignition – the symbolic core of the Australian system – seemed only another massive intrusion on an abused natural scene, a kind of colossal vandalism. Even within the fire community, critics, led mostly by academic biologists, argued for major revisions in the practices and the research agenda of the Australian system. Not fire protection per se, they held, but overall ecological management ought to be the object of fire programmes. Intellectuals pleaded that fire be used to shape a spectrum of desired biotas, not simply to reduce fuel loads according to preset timetables. Thoughtful critics wanted a micro-management of exotic niches and diverse fuels; this demanded a spectrum of fire intensities and frequencies. The public clamored for better, high-technology fire suppression on the 'North American' model. And, an ultimate affront, the public and its elected officials even demanded that the bushfire community scrap their talismanic 'fire bombers' – denounced by critics as 'aeroburning' – and replace them with North American air tankers.[40]

Beyond the ideology was a massive restructuring of the Australian rural scene. Instead of extensive rural usage, more intensive, specialized land usage became the norm. Rural and crown lands were parcelled out to new purposes – to Aboriginal reserves, to suburbs reaching like tentacles from metropolitan city-states, to nature reserves, floral and faunal conservation sites, or components of a national parks and wildlife system.

All this profoundly challenged the assumptions behind the Australian strategy. The strategy had successfully rationalized and redirected folk practices. But it had also expected that that folk heritage would continue to underwrite its popular support. The rural decline, or rather redirection, eroded much of that implicit support. Rightly or wrongly, too, bushfire protection shared in the general popular decline for public forestry. Even the pivotal precept of the strategy – that controlled burning for hazard reduction was the only sensible basis for fire management in Australia – was revealed as less a truth than a truism. It was a tautology that could only be made true to the extent that it could be practised. It was not that the logic of the system was flawed but that it demanded a particular environmental and social context before it could be realised. The system had been conceived and promulgated by native-born Australians who had weathered the 'two fires', but it would have to be accepted by a generation of New Australians who had not. Australia was, in effect, being recolonized by a new set of fire practices.

These would demand a revised science. Bushfire research had been designed to justify the precepts of the Australian strategy and to sustain its implementation in the field. By the late 1970s it showed the same strains. With the retirement of McArthur, bushfire research lost its great patron and the institutional niche that had lofted it into a programme of international significance. What McArthur had brought together now fell apart. The research cadre splintered. Some members joined McArthur in retirement, some (like King) were reassigned to other CSIRO divisions, and some took up posts with the States or such institutions as the Australian Counter-Disaster College. Only one researcher, N.P. Cheney, a McArthur protégé, remained with the Division of Forest Research. At just the time when the U.S. Forest Service was revising its national fire policy to tolerate – even encourage – more prescribed fire and busily restructuring its fire research establishment to reflect that need and become more responsive to its field clientele, Australia was dismantling its own programme, under pressure to resemble more closely the increasingly discredited American model. Each nation was being told to look more like the other.

But there was a conceptual as well as an institutional issue at stake. Critics wanted to reorientate the agenda of bushfire research to include the biology of fire – not simply the physics of how fire spreads or how controlled burning lessens fire intensity, but how fire functions in Australian ecosystems and how broad-acre burning influences the dynamics and structure of those biotas. They wanted fire research to shift its attention from an obsession with bushfire protection to the question of how fire could be regulated to promote larger ecological and social values. Not broad-acre treatments, applicable on a continental scale, but specialized prescriptions for unique environments were the points of their concern.[41] This posed a methodological crisis, for existing empirical relationships could not be extrapolated into the whole spectrum of new conditions, and it would be prohibitively expensive to subject each fuel array and each fire intensity to the same kind of field experimentation that had been conducted historically. What was needed was something like what American researchers had developed – fire behaviour models based on analytical principles that could be applied over a wide range of circumstances, modified by selective field and laboratory testing.

The debate was, in any event, only partly scientific. The burning bush had become an oracle in the restless search for an Australian identity. What was being argued was the relationship of Australians to the bush. Discussions about how Europeans 'adapted' was a coded language for asking about the character of Australians and the legitimacy of Australian society. In this context, technical arguments about fire practices were often irrelevant. McArthur freely confessed to early sceptics that burns were far from uniform, that up to 30% of the targeted landscape might escape any meaningful degree of burning. Tirelessly, he pointed out that foresters were only doing what had been done for tens of millenia in Australia, that they were perpetuating an ecological process of great antiquity. He also brought to public discussion the sense of someone who had spent decades in practical fire control. His more intellectual critics had not, as he had, been charged with the responsibility of ordering volunteer brigades against high-intensity bushfires. To McArthur the issue was both pragmatic and moral: any alternative strategy simply could not work under Australian conditions and it would recklessly endanger volunteer firefighters. Over and again he found himself lecturing Australians about their own continent. Like most of the Australian fire community, he greeted critics with disbelief and a sense of betrayal.[42]

A change in values and expectations had occurred, and it was not enough to protect people and property from fire – or at least not through the old methods. What McArthur considered as a rationalization of indigenous practices, critics perceived as yet another broad-acre manipulation of the rural bush, no different ideologically from the transformation of downs into sheep runs or the indiscriminate clearing of scrub for wheat fields. In particular, routine broad-acre burning seemed to be socially unacceptable for two new environments of great significance to the New Australians – the native bush reserved for indigenous flora and fauna, and the urban bush increasingly developed for affluent ex-urbanites. Here questions of technique evaporated into more ethereal questions of identity, with human values that resisted codification into field meters.

The institutional base for the Australian system dissolved after McArthur resigned, having endured long years of diabetes and kidney dialysis. The old guard retired, were dispersed into other CSIRO divisions, or were reassigned to new forestry-based projects. The great Victorian fires of 1977 inspired another round of political inquiries.[43] When Prime Minister Malcolm Fraser took personal interest in the bushfire question, it was a politically mixed blessing. Fraser wanted to reinvest in bushfire research but out of a desire to bring air tankers to Australia.

CSIRO's Project Aquarius (1981–1984), headed by Cheney, took a more liberal view of its role, reviewing the spectrum of possible bushfire control measures. It seized the opportunity to reexamine and refine much of the data underwriting fire danger meters and to upgrade the sophistication of its analysis. Yet this was a project, not an institute: the programme would dissolve when its originating questions were answered. Before Aquarius concluded its experimental burns in Western Australia, however, South Australia and Victoria were savaged by the worst outbreak of fires since 1939.

THE BURNING BUSH

> Mr. CHYNOWETH – It is very hard for a man to play nature, is it not?
> Mr. PACKHAM – Yes.
> – Parliamentary Hearings on the Ash Wednesday fires (1983)

The Ash Wednesday holocaust of 1983 dramatized the terms of the contemporary debate about Australian bushfires, controlled and wild.[44] There were a 50-year, maybe a 100-year fire. Seventy-seven people died. The fires incinerated 2,528 homes and five sawmills. They burned over two million hectares of forests, native and planted – about 5% of Australia's forested land. Nearly 25% of South Australia's pine plantations were lost in an afternoon. The fires catalyzed another Parliamentary inquiry, this time taking as its charge the Commonwealth as a whole, not a particular State. The fires galvanized the forestry and fire communities. They dramatized, as nothing since the 1939 conflagration and Judge Stretton's forceful prose had, the endemic character of fire in Australia, the degree to which natural evolution and human history had come to share, in Australia, a special relationship to fire. The Ash Wednesday conflagrations brought to a boiling point the debate about the nature of Australians' adaptation to their environment, both the bush into which they intruded and the bush they were actively creating. The new bushman – the colonizer of the urban bush, a tourist-swagman of the wild bush – would have to work out new terms of reconciliation.

Reforms, however, have been modest. CSIRO established a National Bushfire Research Unit (NBRU) under N.P. Cheney, who set about updating the McArthur research agenda, outfitting it with some high-tech. hardware, and reaffirming its political strengths by working with a consortium of State and rural fire services and other CSIRO units. In addition to the traditional themes, the new agenda shifted attention to the urban bush, with a series of inquiries into what caused houses (and people) to survive or perish in major conflagrations. As State forestry agencies were either assigned jurisdiction over nature conservancies or instructed to work with park and wildlife bureaus, more attention was paid to the special environments of Australia for which broad-acre treatments were probably not suitable. Bushfire research broadened its institutional base but lost its intensity; what had been a river became a desert lake, shallow and flooded with temporary run-off from the Ash Wednesday watershed.[45]

The American strategy, too, had fallen into disarray. Battered by environmentalist critics, forestry had sagged in public esteem, and the wildland fire establishment had reluctantly organized itself around the question of wilderness fire. Forest Service hegemony fractured. Land management, not fire protection, formed the basis of new institutional coalitions. New participants brought their own research agendas. Then wilderness fire as an informing problem began dying on its feet. Fire research struggled to survive. One fire laboratory was shut down, another was rededicated for systems analysis. Having moved beyond physics into the biology of fire, researchers now lurched into fire economics, an attempt of justify budgets. American fire research wasted away like an aging weightlifter. Because the Federal government, through the Forest Service, had claimed special responsibilities for fire research, no alternative research centres had developed. Without State support or university bases, no other institution could step into the vacuum. When the nuclear winter controversy rekindled interest in free-burning fire, the wildland fire community was able to participate only marginally.[46]

In all this America and Australia were, once again, friendly antipodes. Australians have to move into wilderness fire management, which the American fire community is leaving, while Americans are preparing to advance into the ex-urban fire scene, complete with volunteer fire brigades, of which Australians have had long experience. Australian bushfire research is fragmenting, unable to transcend an excessive parochialism. American wildland fire research is shrinking, unable to break out of its excessive reliance on the Federal establishment and its cycle of boom and bust. When Americans and Australians talk fire, an exercise in international mateship, they speak always with great good-will and almost always they speak past one another. Each side continues to see the world through the logical prism of its own system. Their fire histories are a parable of Australian-American relations.

Less obviously, perhaps, its fire history may also serve as an historical parable of the New Australia. At issue is not only the reconciliation of New Australian with Old Australia, a problem internal to the history of an unusual continent settled in unusual ways. Those experiments are also part of a global epic of humanity and fire; and as occupants of a notoriously fire-prone continent, Australians have pioneered new technologies, a new strategy, new institutions to relate humans to fire. The Australian strategy has had, and will continue to have, important international consequences. The reciprocity that has always existed between *Homo* and fire was, in Australia, reformed, distilled and magnified.

In his novel *The Tree of Man*, Patrick White describes a bushfire that ravages the countryside. It is a baptism by fire, the successor to the floods that earlier baptized the landscape in water. More properly, it is a day of judgment that tries the character of the community. The fire becomes an occasion for redefining its identity. 'The fact was, the fighters had become not only exhausted but fascinated by the fire ... There were very few who did not succumb to the spell of the fire. They were swayed by it, instead of it by them'. Weeks afterwards as green reclaimed the landscape, the people were 'full of resolutions. Because they had looked into the fire, and seen what you do see, they could rearrange their lives. So they felt'.

So they deserved to feel. When Manning Clark admonished contemporary Australians for living in an 'age of ruins', without unique vision and with only mindless materialism as a surrogate for morality, he was only half-right.[47] He was looking at the shadows when he should have looked at the illuminating fire. The burning bush tells a different story.

Arizona State University, West
U.S.A.

NOTES

* A National Science Foundation grant (SES) supported travel to Australia during June and July, 1986. I wish to thank Mr. N.P. Cheney for his indispensable assistance in making that tour a success.

1 Compilations of historic fires, including Black Thursday, can be found in J.C. Foley, *A Study of Meteorological Conditions Associated with Bush and Grass Fires and Fire Protection Strategy in Australia*, Bulletin No. 38, Bureau of Meteorology (Melbourne, 1947); R.H. Luke and A.G. McArthur, *Bushfires in Australia* (Canberra, 1978); Ted Foster, *Bushfire* (Sydney, 1976); and N.P. Cheney, 'Bushfire Disasters in Australia, 1945–1975', *Australian Forestry*, 39(4) (1976), 245–268.

2 Leonard E.B. Stretton, *The Causes of and Measures Taken to Prevent the Bush Fires of January, 1939 and to Protect Life and Property*, Report of Royal Commission (Melbourne, 1939).

3 Quoted in Michael Dunn, *Australia and the Empire: From 1789 to the Present* (Sydney, 1984), p. 141.

4 My sense of the 'American invasion' comes from Richard White, '"Combating Cultural Aggression"': Australian Opposition to Americanisation', *Meanjin*, 39(3) (October 1980): Michael Dunn, op. cit. (n. 3); and Norman Bartlett, *Australia and America Through 200 Years* (Sydney, 1976).

5 For the floral history of Australia, see J.M.B. Smith, *A History of Australasian Vegetation* (Sydney, 1982); *Evolution of the Flora and Fauna of Arid Australia*, eds. W.R. Barker and P.J.M. Greenslade (Sydney, 1982); *Ecological Biogeography of Australia*, 3 vols., ed. A. Keast (The Hague, 1982); and Mary White, *The Greening of Gondwana* (Sydney, 1986).

The literature on Australian fire is vast. Useful compendia include A. Malcolm Gill and I.R. Noble, *Bibliography of Fire Ecology in Australia* (Sydney, 1986), and *Fire and the Australian Biota*, ed. A. Malcolm Gill (Canberra, 1981).

6 Sylvia Hallam, *Fire and Hearth: A Study of Aboriginal Usage and European Usurpation in Southwestern Australia* (Canberra, 1979), p. vii.

7 Phyllis Nicholson, 'Fire and the Australian Aborigine – an enigma', in *Fire and the Australian Biota*, ed. A. Malcolm Gill et al., p. 61.

[8] Rhys Jones, 'Fire-Stick Farming', *Australian Natural History*, 16(1969), 224–228. For a critical reply, see D.R. Horton, 'The Burning Question: Aborigines, Fire, and Australian Ecosystems', *Mankind*, 13(3) (1982), 237–251. A useful summary of contemporary research is contained in articles by Julian Ford, Sylvia Hallam, and Henry T. Lewis in *Fire Ecology and Management in Western Australian Ecosystems*, ed. Julian Ford, WAIT Environmental Studies Group Report No. 14 (Perth, 1986).
[9] E.H.F. Swain, 'Rural Fires', *Australian Forestry Journal* (March 15, 1927), p. 69.
[10] A.M. Gill, 'Post-settlement Fire History in Victorian Landscapes', in *Fire and the Australian Biota*, pp. 77–98.
[11] Accounts of the 1851 and 1898 fires can be found in Gill, *ibid.* The Gippsland fires are the subject of several features in *Land of the Lyrebird*, ed. Committee of the South Gippsland Pioneers Association (Melbourne, 1920). Virtually all accounts of life in Victoria in 1851 refer to the fires of that year.
[12] For the Australian as Bushman syndrome, see Richard White, *Inventing Australia* (Sydney, 1981); Francis Adams, *The Australians: A Social Sketch* (London, 1893); and Vance Palmer, *The Legend of the Nineties* (Melbourne, 1954).
[13] The American fire literature, particularly scientific, is enormous. Entry can be gained through Stephen Pyne, *Fire in America: A Cultural History of Wildland and Rural Fire* (Princeton, 1982) and *Introduction to Wildland Fire: Fire Management in the United States* (New York, 1984), and H.A. Wright and A.W. Bailey, *Fire Ecology in the United States and Canada* (New York, 1982).
[14] From the Plumas Boundary Report of 1904, quoted in Louis Barrett, 'A Record of Forest and Field in California from the Days of the Early Explorers to the Creation of the Forest Reserves', unpublished U.S. Forest Service report (1935), p. 48.
[15] There are some interesting exceptions. Saltbrush burns rarely and shows little resilience to fire. Brigalow is virtually incombustible except through felling and deliberate firing. See assorted articles in *Management of Australia's Rangelands*, ed. G.N. Harrington et al. (Melbourne, 1984).
[16] See G.C. Stocker and J.J. Mott, 'Fire in the Tropical Forests and Woodlands of Northern Australia', in *Fire and the Australian Biota*, pp. 427–442; *Ecology of the Wet-Dry Tropics*, Proceedings of the Ecological Society of Australia, vol. 13, ed. M.G. Ridpath and L.K. Corbett (Darwin, 1985); and *Proceedings of the Tropical and Arid Fire Symposium*, chair R.H. Luke (Darwin, 1971).
[17] The best survey of the interior is contained in *Management of Australia's Rangelands* op. cit. (n. 15). See also G.F. Griffin et al., 'Wildfires in the Central Australian Rangelands, 1970–1980', *Journal of Environmental Management*, 17 (1983), 311–323. An account of the 1974–75 fire season is available in Luke and McArthur, *Bushfires in Australian*, pp. 339–344.
[18] R.G. Vines, 'A survey of forest fire danger in Victoria (1937–1969)', *Australian Forest Research*, 4(2) (1969), p. 40. This should be compared with Cheney, op. cit. (n. 1).
[19] The most accessible summary of fire weather is in Luke and McArthur, *Bushfires in Australia*. This can be supplemented by *Proceedings of Fire Weather Services Conference, Adelaide, May 1985* (Melbourne, 1986). Most case studies of large fires include meteorological analyses.
[20] Data from J.B. Johnston et al., *Fire Protection and Fuel-Reduction Burning in*

Victoria. Report to the Minister of Forests by a Task Force of Officers of the Forests Commission Victoria (1982), p. 2.

21 See L.T. Carron, *A History of Forestry in Australia* (Canberra, 1985). Compare this to the American experience: Harold Steen, *The U.S. Forest Service: A History* (Seattle, 1976) and Henry Clepper, *Professional Forestry in the United States* (Baltimore, 1976). As an example of foresters as exporters of firefighting, see E.O. Shebbeare, 'Fire Protection and Fire Control in India', *Third Imperial Forestry Conference, Canberra, 1927* (Canberra, 1928).

22 D.H. Hutchins, *A Discussion of Australian Forestry* ... (Perth, 1916). The book is a voluminous and rambling, but fascinating, survey of Australian forestry and fire practices at the time.

23 Sketches of Lane-Poole and Kessell are contained in Athol Meyers, *The Forester* (Hobart, 1985). For Kessell's early views, see his *A Primer on Forestry* (Perth, 1925).

24 G.J. Rodger, *Report of the Royal Commission ... Bush Fires of December 1960 and January, February and March 1961 in Western Australia* (Perth, 1961).

25 Biographical material on McArthur is distressingly light. Basic facts are contained in 'Obituary, Alan G. McArthur', *Australian Forestry*, 41(1978), pp. 189–190. I have supplemented this outline by conversations with N.P. Cheney and R.H. Luke.

26 For the research outside forestry, see R.G. Vines, 'Bushfire Research in CSIRO', *Search* 6(3) (March 1975), pp. 74–79.

27 Basic biographical materials on Luke are available in 'N.W. Jolly Medal, 1980 Award: Robert Henry Luke', *Australian Forestry*, 44(2) (1981). These have been supplemented by conversations between Luke and the author.

28 A reasonably complete survey of American fire policy and programmes is contained in Pyne, *Fire in America*; a new preface, updating the story, graces the paperback edition (1988). Shorter, more recent examinations include Pyne, 'Fire Policy and Fire Research in the U.S. Forest Service', *Journal of Forest History* 25(2) (April 1981), 64–77; 'Vestal Fires and Virgin Lands: A Historical Perspective on Fire and Wilderness', in *Proceedings – Symposium and Workshop on Wilderness Fire*, ed. James Lotan et al., U.S. Forest Service General Technical Report INT-182 (1985), p. 254–266; and 'The Fire Next Time: Old Fire Problems and New Problem Fires', *Ninth Fire and Forest Meteorology Conference* (American Meteorological Society, 1987), pp. 197–203.

For the context of early years, see Pyne, 'Prelude to Sustained-Yield Forestry: The Origins of Systematic Fire Protection in the United States', *History of Sustained-Yield Forestry: A Symposium*, ed. Harold K. Steen (Durham, North Carolina, 1984), pp. 68–78.

29 An abridged survey of the controversy is given in Pyne, ibid. A more detailed recounting can be found in Raymond Clar, *California Government and Forestry*, 2 vols. (Sacramento, 1959, 1969). Also useful are F.E. Olmstead, *Light Burning in the California Forests* (U.S. Forest Service, 1911) and Ashley Schiff, *Fire and Water: Scientific Heresy in the U.S. Forest Service* (Cambridge, Mass., 1962).

30 The full citation should read Coert duBois, *Systematic Fire Protection in the California Forests* (U.S. Forest Service, 1914).

31 See Donald Bruce, 'Light Burning: Report of the California Forestry Committee', *Journal of Forestry*, 21 (February 1923), 129–133.

[32] The importance of fire control to Federal-State programmes is sketched in Pyne, *Fire in America*; Earl Pierce and William Stahl, 'Cooperative Forest Fire Control: A History of Its Origins and Development Under the Weeks and Clarke-McNary Acts' (U.S. Forest Service, 1964); and William Robbins, *American Forestry: A History of National, State and Private Cooperation* (Lincoln, Nebraska, 1985).

[33] William Greeley, 'Better Methods of Fire Control', *Proceedings, Society of American Foresters*, 6 (1911), 165.

[34] In addition to the material in Pyne, *Fire in America*, see Ashley Schiff, op. cit. (n. 29). For a biographical profile of an early forester working in fire research, see Charles E. Hardy, 'The Gisborne Era of Forest Fire Research' (U.S. Forest Service, 1977).

[35] The Flambeau and Euroka experiments grew out of the Cuban missile crisis. For the Australian contribution, see The Technical Cooperation Program, *Report on Mass Fire Symposium, Canberra, February 1969*, 2 vols. (Defense Documentation Center, 1970).

[36] The Tall Timbers Research Station sponsored fire ecology symposia from 1962 to 1976 and privately published the papers each year as a volume. An overview is contained in E.V. Komarek, *Tall Timbers Research Station: A Quest for Ecological Understanding*, Misc. Publication No. 5 (Tallahassee, Florida, 1977). As a forum for Australian sentiments, see, for example, A.B. Mount, 'An Australian's Impression of North American Attitudes to Fire', *Proceedings, Annual Tall Timbers Fire Ecology Conference*, No. 9 (Tallahassee, Florida, 1969), pp. 109–118. The fire study tour reports written by Australians are equally revealing.

[37] The critical debate about fire bombers and water bombers was conducted during the 1960s and is reported in *Application of Aircraft in Bushfire Control: Royal Aeronautical Society Symposium* (Royal Aeronautical Society, 1969). The best summary of contemporary Australian sentiment about air tankers is contained within I.T. Loane and J.S. Gould, *Aerial Suppression of Bushfires: Cost-Benefit Study for Victoria* (CSIRO National Bushfire Research Unit, 1986).

[38] My sources for Aussie demographics are Geoffrey Sherington, *Australia's Immigrants* (Sydney, 1980); I.H. Burnley, *Population, Society, and Environment in Australia* (Melbourne, 1982); and *Urbanization in Australia: The Post-War Experience*, ed. I.H. Burnley (Melbourne, 1972).

[39] The literature on environmental activism in Australia is growing rapidly. Two historical surveys are Geoffrey Bolton, *Spoils and Spoilers* (Sydney, 1981) and Derek Whitelock, *Conquest to Conservation* (Adelaide, 1985).

[40] The numerous symposia on fire ecology all include discussions about policy. A more or less comprehensive survey is contained in House of Representatives, Standing Committee on Environment and Conservation, *Environmental Impact of Bushfires*, Hansard report, 13 vols. (Canberra, 1984). A vigorous if unfair critique of aero-burning can be found in Australian Conservation Foundation, *Bushfire Control and Conservation*. Viewpoint Series No. 5 (Sydney, 1970).

[41] The best summary of contemporary concerns is contained in Australian Forestry Council, *Australian Bushfire Research: Background, Guidelines, and Directory* (Canberra, 1987). A thoughtful critique of research agendas established too exclusively by forestry is available in A. Malcolm Gill, *Research for the Fire Manage-*

ment of Western Australian State Forests and Conservation Reserves (Perth, 1986).

[42] Among useful summaries of McArthur's attitudes are 'The Historical Place of Fire in the Australian Environment', *Second Fire Ecology Symposium*, Monash University (Melbourne, 1970); 'Prescribed Burning in Australian Fire Control', *Australian Forestry*, 30(1) (March 1966), 4–11; and his influential book, co-authored with R.H. Luke, *Bushfires in Australia* (Canberra, 1978).

[43] Something of the fires and their impact can be found in A.G. McArthur et al., *The Fires of 12 February 1977 in the Western District of Victoria* (Melbourne, 1982) and E.H.E. Barber, *Report of the Board of Inquiry into the Occurrence of Bush and Grass Fires in Victoria* (Melbourne, 1977).

[44] The literature on the Ash Wednesday fires is already large. The best summary is contained in the Parliamentary hearings, *Environmental Impact of Bushfires*, 13 vols.

[45] For a sense of contemporary tendencies to fragmentation, refer to Australian Forstry Council, op. cit. (n. 41).

[46] The American fire research scene, as it existed just prior to the nuclear winter controversy, is reviewed in Pyne, 'Recent History of Fire Control and Use; or, The Knowing Flame', *Foresters' Future: Leaders or Followers?*, Proceedings of the 1985 Society of American Foresters National Convention (Washington, D.C. 1986), pp. 69–80.

[47] Manning Clark, *A Short History of Australia*, 2nd rev. ed. (New American Library, 1980), p. 269.

NOTES ON CONTRIBUTORS

DAVID WADE CHAMBERS completed his PhD at Harvard in the history of science and since 1977 has been Chair of the Open Campus Program in Social Studies of Science at Deakin University. He is the author of a number of textbooks, including *Red and Expert*, on Chinese science and technology during the Cultural Revolution, and *Singing the Land, Signing the Land* (with Helen Watson), comparing Australian Aboriginal and European understandings of nature. He is currently President of the Australasian Association for the History, Philosophy and Social Studies of Science.

C. STEWART GILLMOR is Professor of History and Science at Wesleyan University, Connecticut. He was educated in electrical engineering at Stanford and in history and philosophy of science at Princeton. His first job was as ionospheric physicist with the Sixth Soviet Antarctic Expedition, 1960–62. He has published on ionospheric physics and on the history of physics and engineering, and is editor of the series *The History of Geophysics*. He is Secretary/Treasurer of the History of Physics Division of the American Physical Society and is a sometime professional jazz musician.

DAVID KNIGHT has taught history of science in the University of Durham since 1964. From 1981 to 1988 he was editor of the *British Journal for the History of Science* and he is now General Editor of the Blackwell's Scientific Biography Series. His book *The Age of Science* is concerned with the development of science in the nineteenth century.

HOMER LE GRAND is in the Department of History and Philosophy of Science at the University of Melbourne. A previous life was devoted to technical studies of eighteenth-century chemistry. His current research in modern geology and geophysics draws upon historical, philosophical and sociological approaches to issues of scientific change and theory choice.

R.W. Home and S.G. Kohlstedt (eds.), International Science and National Scientific Identity, 289–291.

Recent publications in this vein include *Drifting Continents and Shifting Theories* (1988).

ELIZABETH DALTON NEWLAND is a graduate of the Universities of Sydney and New South Wales. She is a trained librarian and for a number of years was Librarian at the Australian Atomic Energy Commission, Lucas Heights. She is currently researching the life and research activities of Dr. George Bennett for a PhD at the University of Wollongong.

MICHAEL A. OSBORNE studied biology and political science before taking his PhD in the history of science at the University of Wisconsin. Currently a lecturer in the History Department, University of California, Santa Barbara, he was previously a research associate with the Centre de Recherche en Histoire des Sciences et des Techniques in Paris. A historian of science and medicine in Europe and the European empires, his book, *The Société zoologique d'acclimatation and the New French Empire: Science and Political Economy during the Second Empire and Third Republic*, is forthcoming with Indiana University Press.

STEPHEN J. PYNE is a professor in the American Studies programme at Arizona State University (West Campus). His historical interests extend to exploration, natural science, and the environment, best expressed in *The Ice: A Journey to Antarctica* (1986). He has written several books on fire, including *Fire in America* (Princeton, 1982), *Introduction to Wildland Fire* (1984), and *Burning Bush: A Fire History of Australia* (1990).

JAN SAPP is Senior Lecturer in the Department of History and Philosophy of Science at the University of Melbourne. He has written extensively on the history of the modern life sciences. He is author of *Beyond the Gene: Cytoplasmic Inheritance and the Struggle for Authority in Genetics* (1987) and *Where the Truth Lies: Franz Moewus and the Origins of Molecular Biology*(1990). He is at present writing about symbiosis in relation to cell theory and evolutionary innovation.

ROBERT A. STAFFORD earned B.A. and M.A. degrees in history at the University of Washington in Seattle and a D.Phil. in modern history at Oxford University. His dissertation, on the role of the geologist Sir Roderick Murchison in promoting the scientific exploration of the British empire during the nineteenth century, was published in 1989 by Cambridge University Press under the title *Scientist of Empire*. He has held post-doctoral research fellowships in the Department of History and Philosophy of Science at the University of Melbourne and the Department of History at La Trobe University. He is currently employed as a technical writer in Melbourne.

The Editors

R.W. HOME studied physics and then history and philosophy of science at the University of Melbourne before completing a PhD in history and philosophy of science at Indiana University. He has been Professor of History and Philosophy of Science at the University of Melbourne since 1975. He has published extensively on the history of eighteenth-century physics and, more recently, on the history of science in Australia. He is editor of *Historical Records of Australian Science* and is currently preparing a history of the Australian physics community to 1945.

SALLY GREGORY KOHLSTEDT is Professor of the History of Science at the University of Minnesota. She has written on the institutional history of science with particular reference to the American Association for the Advancement of Science, has investigated the participation of women in science and technology, and has an interest in national comparative and co-operative aspects of science. She is currently editing selected essays of George Brown Goode and working on the history of natural history museums in the United States.

INDEX*

NOTE

* This index was compiled with the assistance of Horace Taft-Ferguson, a graduate student in the History of Science and Technology Program at the University of Minnesota.

AUSTRALASIAN STUDIES IN HISTORY AND PHILOSOPHY OF SCIENCE

General Editor:
R. W. Home, *University of Melbourne*

Publications:

1. R. McLaughlin (ed.): *What? Where? When? Why?* Essays on Induction, Space and Time, Explanation. Inspired by the Work of Wesley C. Salmon. 1982 ISBN 90-277-1337-5
2. D. Oldroyd and I. Langham (eds.): *The Wider Domain of Evolutionary Thought.* 1983 ISBN 90-277-1477-0
3. R. W. Home (ed.): *Science under Scrutinity.* The Place of History and Philosophy of Science. 1983 ISBN 90-277-1602-1
4. J. A. Schuster and R. R. Yeo (eds.): *The Politics and Rhetoric of Scientific Method.* Historical Studies. 1986 ISBN 90-277-2152-1
5. J. Forge (ed.): *Measurement, Realism and Objectivity.* Essays on Measurement in the Social and Physical Science. 1987 ISBN 90-277-2542-X
6. R. Nola (ed.): *Relativism and Realism in Science.* 1988 ISBN 90-277-2647-7
7. P. Slezak and W. R. Albury (eds.): *Computers, Brains and Minds.* Essays in Cognitive Science. 1989 ISBN 90-277-2759-7
8. H. E. Le Grand (ed.), *Experimental Inquiries.* Historical, Philosophical and Social Studies of Experimentation in Science. 1990 ISBN 0-7923-0790-9

KLUWER ACADEMIC PUBLISHERS – DORDRECHT / BOSTON / LONDON

www.ingramcontent.com/pod-product-compliance
Ingram Content Group UK Ltd.
Pitfield, Milton Keynes, MK11 3LW, UK
UKHW040228290926
15471UKWH00001B/65
* 9 7 8 9 4 0 1 1 3 7 8 7 4 *